Springer Series on
SIGNALS AND COMMUNICATION TECHNOLOGY

Signals and Communication Technology

Distributed Cooperative Laboratories
Networking, Instrumentation, and Measurements
F. Davoli, S. Palazzo and S. Zappatore (Eds.)
ISBN 0-387-29811-8

The Variational Bayes Method in Signal Processing
V. Šmídl and A. Quinn
ISBN 3-540-28819-8

Topics in Acoustic Echo and Noise Control
Selected Methods for the Cancellation of Acoustical Echoes, the Reduction of Background Noise, and Speech Processing
E. Hänsler and G. Schmidt (Eds.)
ISBN 3-540-33212-x

EM Modeling of Antennas and RF Components for Wireless Communication Systems
F. Gustrau, D. Manteuffel
ISBN 3-540-28614-4

Interactive Video Methods and Applications
R. I Hammoud (Ed.)
ISBN 3-540-33214-6

ContinuousTime Signals
Y. Shmaliy
ISBN 1-4020-4817-3

Voice and Speech Quality Perception
Assessment and Evaluation
U. Jekosch
ISBN 3-540-24095-0

Advanced ManMachine Interaction
Fundamentals and Implementation
K.-F. Kraiss
ISBN 3-540-30618-8

Orthogonal Frequency Division Multiplexing for Wireless Communications
Y. (Geoffrey) Li and G.L. Stüber (Eds.)
ISBN 0-387-29095-8

Circuits and Systems Based on Delta Modulation
Linear, Nonlinear and Mixed Mode Processing
D.G. Zrilic ISBN 3-540-23751-8

Functional Structures in Networks
AMLn—A Language for Model Driven Development of Telecom Systems
T. Muth ISBN 3-540-22545-5

RadioWave Propagation for Telecommunication Applications
H. Sizun ISBN 3-540-40758-8

Electronic Noise and Interfering Signals
Principles and Applications
G. Vasilescu ISBN 3-540-40741-3

DVB
The Family of International Standards for Digital Video Broadcasting, 2nd ed.
U. Reimers ISBN 3-540-43545-X

Digital Interactive TV and Metadata
Future Broadcast Multimedia
A. Lugmayr, S. Niiranen, and S. Kalli
ISBN 3-387-20843-7

Adaptive Antenna Arrays
Trends and Applications
S. Chandran (Ed.) ISBN 3-540-20199-8

Digital Signal Processing with Field Programmable Gate Arrays
U. Meyer-Baese ISBN 3-540-21119-5

Neuro-Fuzzy and Fuzzy Neural Applications in Telecommunications
P. Stavroulakis (Ed.) ISBN 3-540-40759-6

SDMA for Multipath Wireless Channels
Limiting Characteristics and Stochastic Models
I.P. Kovalyov ISBN 3-540-40225-X

Digital Television
A Practical Guide for Engineers
W. Fischer ISBN 3-540-01155-2

Speech Enhancement
J. Benesty (Ed.)
ISBN 3-540-24039-X

Multimedia Communication Technology
Representation, Transmission and Identification of Multimedia Signals
J.R. Ohm ISBN 3-540-01249-4

Information Measures
Information and its Description in Science and Engineering
C. Arndt ISBN 3-540-40855-X

Processing of SAR Data
Fundamentals, Signal Processing, Interferometry
A. Hein ISBN 3-540-05043-4

continued after index

MULTIMEDIA DATABASE RETRIEVAL: A HUMAN-CENTERED APPROACH

PAISARN MUNEESAWANG
College of Information Technology
United Arab Emirates University
Al-Ain, United Arab Emirates

LING GUAN
Department of Electrical and Computer Engieering
Ryerson University
Toronto, Canada

Springer
Boston/Dordrecht/London

Paisarn Muneesawang
College of Information Technology
United Arab Emirates University
Al-Ain, United Arab Emirates

Ling Guan
Department of Electrical and Computer Engineering
Ryerson University
Toronto, Canada

Multimedia Database Retrieval: A Human-Centered Approach

e-ISBN 0-387-34629-5

ISBN: 978-1-4419-3815-2 **e-ISBN: 978-0-387-34629-8**

Printed on acid-free paper.

9 8 7 6 5 4 3 2 1

springer.com

Contents

Acknowledgments vii

1. INTRODUCTION 1
   1 The Human Factor in Multimedia Search and Retrieval 2
   2 Human-Centered Search and Retrieval Techniques 3
   3 Techniques Presented in This Book 6
   4 Applications Included in This Book 9
   5 Organization of This Book 10

2. INTERACTIVE CONTENT-BASED IMAGE RETRIEVAL 11
   1 Introduction 11
   2 Interactive Framework 13
   3 Radial Basis Function (RBF)-Based Relevance Feedback (RF) 17
   4 Learning and Characterization 19
   5 Application to Texture Image Retrieval 25
   6 Application to Digital Photograph Collection 31
   7 A Computer Aided Referral (CAR) System for Mine Target Detection in Side-Scan Sonar Images 42

3. HIGH PERFORMANCE RETRIEVAL: A NEW MACHINE LEARNING APPROACH 47
   1 Introduction 47
   2 Local Model Networks (LMN) 48
   3 Learning via Local Model Network 51
   4 Machine Learning-Based Interactive Retrieval 54
   5 Adaptive Local Model Network 55

6 Application to Corel's Photograph Collection 59
7 Chapter Summary 66

4. AUTOMATIC RELEVANCE FEEDBACK 75
1 Introduction 75
2 Automatic Interaction by Self-Organizing Neural Network 77
3 Descriptors in JPEG and Wavelet Transform (WT) Domains 84
4 Visual Descriptor for Relevance Classification 88
5 Retrieval of JPEG and Texture Images in Compressed Domain 90
6 Chapter Summary 99

5. AUTOMATIC RETRIEVAL ON P2P NETWORK AND KNOWLEDGE-BASED DATABASES 107
1 Introduction 107
2 Image Retrieval in Distributed Digital Libraries 107
3 Peer-to-Peer (P2P) Retrieval System 109
4 Knowledge-Based Image Retrieval 116
5 Relevance Feedback with Region of Interest 121

6. ADAPTIVE VIDEO INDEXING AND RETRIEVAL 127
1 Introduction 127
2 Framework of Adaptive Video Indexing (AVI) 131
3 Adaptive Video Indexing 132
4 Application to CNN News Video Retrieval 140
5 Automatic Reclusive Video Retrieval 147
6 Chapter Summary 156

7. MOVIE RETRIEVAL USING AUDIO-VISUAL FUSION 159
1 Introduction 159
2 Interactive System for Movie Retrieval 159
3 Audio-Visual Cue 163
4 Chapter Summary 172

References 173

Index 185

Acknowledgments

The first author would like to thank the Royal Thai Government Scholarship, Naresuan University, the Canada Research Chair Program, Canada Foundation for Innovations, and Ryerson University for supporting the research works presented in this book.

The authors would like to thank their colleagues/students in the Signal and Multimedia Processing Group at the University of Sydney, and in Ryerson Multimedia Research Laboratory at Ryerson University, especially Dr Ivan Lee, Dr Jonathan Randall, Dr Jose Lay, Tahir Amin, Howard Lee, Mingmei Diao, Yupeng Li, and Dr Hau San Wong, for their help throughout the course of this research as well as writing this book. They would also like to thank Dr Willie Harris of United Arab Emirates University for all the proofreading and corrections.

Finally, the first author thanks his wife, Sirirut, and his son, Taksaphon, for their love and understanding during this time. He also thanks Ms. Akiko Takesue for help with research.

Acknowledgments

[illegible]

[illegible]

[illegible]

Chapter 1

INTRODUCTION

Multimedia is everywhere. Recent advances in computer software and hardware technology has precipitated a new era in the way people create and store data. Millions of multimedia documents—including images, videos, audio, graphics, and texts—can now be digitized and stored on just a small collection of CD-ROMs. This allows users to group, manipulate and view files with the drag of a mouse. The digitization of multimedia is a key reason for the fast-growing number of digital collections for personal, business and public use. Digital libraries have been established to satisfy individual needs for such diverse applications as the entertainment industry, distance education, telemedicine, and Geographic Information System (GIS). The emergence of internet technologies permits users from different geographic locations to share digital collections without boundaries making the web a universally accessible multimedia library.

As a result of the recent explosion in the quantity of digital media, there is an urgent need for new and better techniques for accessing data. Indexing and retrieval are at the heart of multimedia system design—large amounts of multimedia data may not be useful if there are no effective tools for easy and fast access to the collected information. Once collected, the data must be organized efficiently, so that a query search via a search engine will yield a limited, yet useful, number of results. The retrieval process is designed to obtain limited information that satisfies a user at a particular time and within a particular domain application; however, this does not often work as efficiently as intended. A significant challenge, therefore, is to develop techniques that can "interpret" the multimedia contents in large data collections to obtain all the information items relevant to the user query, while retrieving as few non-relevant ones as possible.

This book presents new techniques of multimedia retrieval, which allows full and efficient use of the multimedia data libraries. The focus is on image and video retrieval, using content-based methods and technologies. The book demonstrates state-of-the-art applications within small, medium, and large databases, in various domains.

1. The Human Factor in Multimedia Search and Retrieval

Visual seeking is the process of communication between a user and a computer system, and requires a decision-making method for both sides. The user expresses his or her information needs in the form of a query image (or video) to a multimedia access system, and expects the system to return the most similar or relevant files. This visual seeking clearly involves interpretations of visual content by both the user and the computer. For successful retrieval, it is necessary that the user and the computer system use the same interpretation criteria.

However, human beings are complex creatures, and their motivations and behaviors are difficult to measure and characterize. As a result, the interpretation criteria utilized by human users are not fixed. Human beings are capable of interpreting and understanding visual contents—to simultaneously synthesize context, form and content—beyond the capability of any current computer method. Human interpretation depends on individual subjectivity and the information needed at a particular time and for a particular event. In addition, users learn from the available information (search results) to recognize their needs and refine their visual information requests (refine the queries) [113]. In other words, the interpretation of visual content made by a human user is non-stationary, or fuzzy, and it is very difficult to describe by fixed rules. A human user is an adaptive-learning component in the decision-making system.

In order to build a computer system to simulate and understand the decision-making processes of human beings, the above-mentioned characteristics of adaptability should be taken into account. Learning to adapt and to optimize decision-making is primary to the goal of creating a better computer-based retrieval system.

The first aim of techniques presented in this book is, therefore, to explore an adaptive machine that can learn from its environment, from both user advice as well as self-adaptation. Specifically, the adaptive machine requires two important properties to achieve this purpose: nonlinear decision-making ability, and the ability to learn from different sources of information (i.e., multi-modeling recognitions). By embedding these two properties into the computer, the system can potentially learn what humans regard as significant. Through the human-computer interaction process, the system will develop the ability to follow non-stationary human decision-making in visual-seeking environments.

The second aim is to optimize the learning system by incorporating self-organizing adaptation into the retrieval process. Optimization is the process of reducing the user's direct input, as well as the process of adjusting the learning system architecture for flexibility in practical use in multimedia retrieval. While user interaction provides a valuable information exchange between the user and the computer, programing the computer system to be self-learning is highly desirable. This book presents the newly developed methods that are intelligent enough to advise the adaptation modules to minimize user input. A retrieval system based on these techniques reduces the user workload in interactive retrieval.

2. Human-Centered Search and Retrieval Techniques

In recent years, *adaptive retrieval* via user-interaction interface has become one of the most active research areas in multimedia retrieval [32, 55, 76, 77]. It has been introduced to address the difficulties faced by the traditional non-adaptive content-based retrieval (CBR) approaches—the gap between low-level visual features used in retrieval and high-level perceptive concepts. The retrieval process typically begins with the submission of a query, and is followed by repeated relevance feedback. Here, the user manages the retrieval system, through selections of his or her information needs, in the form of positive or negative samples. One may think of interactive retrieval as the problem of training machines to identify relevant images in a database given only a "small" set of examples and user supervision. The strategy has brought up many new challenges, and has been used in many modern retrieval systems, including QBIC [1], PicHunter [77], PictoSeek [3], and MARS [8].

Interactive Learning Method

Attempts have been made to integrate relevance feedback (RF) strategies from information retrieval (IR) into image retrieval applications [3, 4, 8, 9, 11]. The integrated RF paradigm incorporates various term-weighting models (TWMs) to characterize images for database indexing. The method implements RF for query reformulation through a process of expanding the terms or expressions of interest. This idea can be further augmented by incorporating a weight adaptation metric in the form of linear and quadratic functions [6, 10, 64–67, 70, 75]. The application is based on the selection of suitable weighting schemes for pattern-similarity ranking. Using this paradigm, different weighting schemes and basis functions have also been explored to improve the functionality of the metric adaptation [68–70]. More recently, the combination of a query reformulation model and an adaptation metric has shown effective learning with respect to the retrieval accuracy and convergence speed of the learning system [14, 71–74]. The fundamental concepts derived from this framework have become significant for many newly-proposed retrieval systems that are directed to a more complex level.

The RF for image retrieval can be treated as a supervised learning paradigm related closely to the pattern-recognition problem, where the applications of machine-learning methods can be suitably implemented. Machine-learning methods, such as neural networks, Bayesian theory, the support vector machine (SVM), and discriminant-expectation maximization (D-EM), are well-known for pattern recognition and classification problems. Consequently, these intelligent machines have been adapted to meet evolving needs in this context [32, 55, 52, 57, 59, 63]. However, the nature of interactive retrieval is notable for its pattern-recognition problems. It is more practical for real-time learning implementation, as well as for dealing with a small volume of training samples. The information available to make a decision is often less than precise. When dealing with user involvement, a preferable interaction method would require few feedback samples. As a result, some of these "direct" applications of machine learning methods, are not well-suited to this particular application, when compared to other pattern-recognition tasks. While these machine-learning approaches display a more complex and sophisticated strategy, compared to the previous adaptive query model and weighted metric, they usually require large volumes of training samples.

Consider the direct application of SVM as an example of the above predicament. It is demonstrated in [63] that SVM requires the user to mark all 100 retrieved samples of each round of interaction to improve the nonlinear classification capability. However, this is not an easy task for the user. In contrast to the previous method, Chen *et al* [32] overcome this problem by modifying SVM to deal with only one class, using RF learning to construct a hyperplane as the direction surface to maximize the margin of separation between positive and negative examples. Here, the assumption is that positive samples can be clustered in a certain way, while negative samples cannot cluster.

In [58–61], neural network models have been adopted because of their learning capability and their ability to simulate universal mapping. Laaksonen *et al* [61], developed PicSom, a system that uses self-organizing feature maps (SOM) to map positive and negative samples into their associated impulses, so that the system is able to isolate only those maps features that provide the most valuable information for each individual query instance. Other supervised network models have been adopted [58, 59, 63] where network weights are trained through a relevance-feedback process. Although the supervised learning networks can achieve good results, a chief disadvantage is the large number of training samples required.

Wu *et al* [56, 57] suggested that a limited number of training samples in RF learning will have poor generalization performances. Thus, to obtain a better similarity measurement, a probabilistic approach is developed to employ both *labeled* (feed-back) images and *unlabeled* images in a given database. The Discriminant-Expectation Maximization (D-EM) method was proposed to perform discriminant analysis in order for the computer to learn the generative model in a lower-dimensional subspace; this relaxes the assumption of the probabilistic structure of the data distribution. In this method, however, the computation can be a concern for large datasets [51]. Vasconcelos *et al* [55] adopted the Bayesian learning machine to implement relevance feedback in order to obtain a mixture of Gaussian models for capturing image regions for matching purposes. Both D-EM and the Bayesian learning machine performed best only if prior knowledge about the data distribution was available. Meanwhile, Wang *et al* [50] adopted an optimal filter to implement RF in which the filter's coefficients were computed by the traditional least mean square (LMS) and recursive least mean square (RLS) algorithms. This provided new types of RF learning. According to the experiments on the Corel test set, the adoption of these filter algorithms gave quite a low retrieval accuracy—only 20% precision after RF learning.

Alternatively, RF can be adopted for off-line learning to create user profiles for further search and retrieval. The general approach for deriving such a profile is based on collecting relevant information from the user, deriving preferences from this information, and modifying the user profile accordingly. Zhuang *et al* [78] has developed a network of semantic templates with the support of WordNet to allow semantic searches via a keyword language. In PicHunter [77], a profile is in the form of a probability distribution that characterizes the probability of each image in the database as a likely target. By its nature, the user profile is relatively easy to use as a surrogate for queries. However, modifications to the user profile should probably be made with caution: a modification made on the basis of a single query session may adversely affect future query sessions due to factors peculiar to that one session.

Adaptive Retrieval for Video

While many RF models have been successfully developed for still-image application, they have not yet been widely implemented for video databases. This is because effective content-analysis through RF learning must also capture spatio-temporal information from a video, and not just spatial information, as required for a single image. The fundamental requirement is a representation that allows RF processes to capture sequential information on a video file. Instead of single-frame content as in still-image applications, complex representation is needed on the different levels of video intervals, such as: shot, scene, and story. There is also another difficulty faced by the practical application of RF learning in this domain. Compared to the image database, the interactive retrieval process on video samples can be a time-consuming task, since video files are usually large. The user has to play all of the retrieved videos in order to make a judgment of relevance. Furthermore, on a network-based database, this application requires a high-bandwidth transmission during the interactive learning process.

Traditionally, video indexing and retrieval have been done by segmentation to obtain video intervals, and have used the content of representative frames, or the motion features within the intervals, for characterization of the video [122, 132–136]. The challenge is to develop high-level retrieval techniques that detect and recognize objects, events, and video captions [121, 126, 130]. The identified objects, events, and video captions are usually annotated with text (terms), and are used for representation of the content. These techniques, however, limit themselves to only a few terms, since the visual contents are often difficult to describe in few words. Furthermore, high-level techniques can be effective only when applied to particular types of video data, and they can usually only support retrieval using key-word search strategies.

A video retrieval strategy presented in this book combines new types of video representation and *automatic* RF to overcome the above challenges. This representation takes into account the spatio-temporal information and allows effective content analysis of video, which can be adopted for representing a video scene or story. Unlike the traditional RF process, the current system implements self-adaptation RF architecture that can be automatically executed with no user input. This technique can avoid the time-consuming task of user-interaction, and allows suitable implementation of video retrieval on the network-based database.

3. Techniques Presented in This Book

To tackle the obstacles in the development of modern multimedia retrieval systems, a combination of machine intelligence and human intelligence is adapted to the work. The book firstly presents a newly developed technique, human-controlled interactive content-based retrieval (HCI-CBR) that provides quantities of user-computer characterizations. It then introduces a machine-controlled interactive content-based retrieval (MCI-CBR) technique to overcome some difficulties introduced by HCI-CBR for practical multimedia retrieval. The book also presents a human/machine controlled adaptation system that is specific to video retrieval. The techniques described in this book can be summarized as follows:

Human-Controlled Interactive Content-Based Retrieval

The development of two HCI-CBR systems are presented. Unlike previously proposed HCI-CBR systems [6, 11, 14, 74], the current adaptive system implements a 'nonlinear' discriminant analysis method to simulate human perception. Traditional relevance feedback (RF) is replaced with a nonlinear model which is composed of an expansion set of the radial-basis functions (RBF) [12, 13] to increase discriminant power in evaluating image similarity. The most important concept in this part of the work is to allow users to communicate their preferences to the adaptive retrieval system, through the provision of qualitative training examples, in the form of selected images in both positive and negative senses; this is a more comfortable way of specifying preferences for humans than merely selecting numerical values for a set of parameters.

This model incorporates and emphasizes various new features not found in earlier interactive retrieval systems. Many of these are imparted by nonlinear discriminant analysis, with a high degree of adaptivity through the associated training algorithm, which uses a small set of positive and negative samples provided by the user to detect all significant image features and improve retrieval accuracy. Compared to previous attempts, this system specializes in effective learning using a very small set of input samples. Since it employs a quick-learning capability, a significant improvement in retrieval precision can be immediately achieved within one to two feedback iterations. These two properties make the current system very attractive for practical implementation. Furthermore, it is demonstrated that this system outperforms, in terms of retrieval accuracy and learning speed, previously proposed HCI-CBR systems [5, 14] that utilize different optimal RF-learning criteria and optimal query designs, including other variations of RF-based learning methods [11, 3, 10].

In the first adaptive system described above, it was observed that the nonlinear model—based on the nonlinear RBF strategy—can present the human-preferred image similarity in a satisfactory way. There are, however, still some inadequacies in characterizing image similarity, primarily due to the local context defined by the current query; this requires more local modeling strategies. This aspect of the similarity-matching problem, which is not explicitly modeled in our previous adaptive system, in turn requires a more local characterization that can fully exploit the local data information. This motivated the development of the Local Model Network (LMN) in the second adaptive system. The mixture of Gaussian models via LMN is used to represent multiple types of model information for the recognition and presentation of images by human beings and machines. Here, the system utilizes multiple types of modeling information to acquire and develop its understanding of the image similarity.

Machine Controlled Interactive Content-Based Retrieval

MCI-CBR is a new challenge for automatic interactive retrieval, as well as the minimization of human involvement. In practical internet-based multimedia applications, it is desirable to reduce the transmission of image/video files between the clients and the service provider by minimizing user feedback and, thus, make better use of the transmission bandwidth and improve processing time. In the MCI-CBR system, a self-organizing adaptive network [35, 36] is derived to *guide* the adaptation of the RF-based learning systems. This can be done in both automatic and semi-automatic fashions. The MCI-CBR architecture is sufficiently flexible for practical implementation on internet-based database, since it incorporates compressed-domain descriptors that allow for direct access to compressed databases in an efficient way.

Minimizing the user's direct participation provides a more user-friendly environment, and reduces errors caused by inconsistent human performance. An important new concept arising from this work is an alternative viewpoint of relevance feedback parameters as self-organizing neuronal weights, which are trainable through the process of *unsupervised competitive learning* [35, 36]. As a result of the self-organizing process, the semiautomatic RF methods can be flexible in the choice of sample training size, thereby achieving better generalized performance from the RF learning machines. This enables the semi-automatic method to reach optimal performance more quickly than direct user controlled RF methods. This is the first time that the theory of interactive retrieval has been combined with the theory of the self-organized learning to provide independent learning for the implementation of adaptive content-based retrieval systems.

MCI-CBR system architecture is dependent upon domain applications. It is suitable with online retrieval systems where multimedia databases are located remotely, such as on peer-to-peer (P2P) network. MCI-CBR offers automatic function that is highly flexible for retrieval processing in various nodes on the P2P network. This increases the system's capability in allocating network resources such as bandwidth and individual workload.

In order to achieve a performance equal to human users, MCI-CBR system needs some levels of *knowledge* to identify relevancy of multimedia content. This knowledge is a pre-defined item that indicates user preferences applied to specific domain application. For example, in photograph collections, "region-of-interest" (ROI) demonstrates the weighting scheme for different regions in a photo, which give a better understanding of the photo's contents. Thus, MCI-CBR can acquire knowledge from ROI to obtain relevant data items that accurately matches user's preferences.

Adaptive Video Indexing and Retrieval

Effective video retrieval requires a representation specifically applied to the time-varying nature of video, rather than the application of still-image techniques. The book presents a framework based on adaptive video indexing (AVI) technique that specific to video database applications discussed as follows:

Fist, AVI technique is applied for video characterization at various levels, specifically by shot, group of shots, or story. Compared with traditional techniques, which use either a representative key-frame or a group of frames, the AVI accomplishes a new view of video descriptors: it can organize and differentiate the importance of the visual contents in various frames, and it describes the contents of the entire video within a shot, group of shots, and a story. This provides accessing video collections from a multi-level perspective. In other words, unlike previous querying methods that were limited to video shots or key-frames [40, 41], AVI offers users who wish to retrieve a video group or story, queries using video clips that contain more accurate narrative.

Second, AVI is applied to human controlled and automatic RF systems. The book describes the method for integration of AVI with a self-training neural network structure to adaptively capture different degrees of visual importance in a video sequence. This network structure implements automatic RF process through its signal propagation with no user input to allow for higher accuracy in retrieval.

Third, an audio-visual fusion that combines visual descriptor with audio descriptor is presented. In this system, multimodality signals of video are used to extract high-level descriptors in order to increase the system's capability for retrieving videos via concept-based queries. For example, it utilizes the concepts "dancing" and "gun shooting" for retrieval of relevant clips from a large collection of movies.

4. Applications Included in This Book

This book is an application oriented reference for multimedia search and retrieval. It deeply focuses on the following applications:

- *Computer-Aided Detection system (CAD).* In Chapter 2, content-based analysis and retrieval techniques are applied to a CAD system for detection of underwater mines in side-scan sonar images.

- *Texture Database Application.* Texture analysis and retrieval has been widely used in various domains, such as (1) in browsing and interpretation of aerial photo in Geography Information System (GIS) and remote sensing; and (2) in analysis and modeling of regions with clustered micro-calcification in a digital mammogram in medicine. In Chapters 2 and 4, CBIR techniques are applied extensively with texture analysis and retrieval.

- *Photograph Database Application.* Indexing and retrieval techniques are designed for collection of photographs which are stored in centralized and de-centralized networked databases, photographs with predefined ROIs, as well as in various standard formats (e.g., JPEG). These are presented in Chapters 2- 4.

- *News Videos Application.* The techniques developed in Chapter 6 are dedicated to indexing and retrieval of CNN news videos.

- *Movie and Audio Databases.* In this domain, retrieval techniques are based on audio and visual content characterizations from multimodality signals, which are presented in Chapter 7.

5. Organization of This Book

This book is organized into 7 chapters. The contents of each chapter are summarized as follows:

Chapter 2 describes the work on the first human-controlled interactive content-based retrieval (HCI-CBR) system. It introduces a nonlinear model using an expansion set of radial basis functions, and the associated learning algorithms. The system performance is then demonstrated in comparison with other recently-developed HCI-CBR systems as well as with a non-interactive system. The applications involved in the experiments are texture retrieval and image retrieval on a general photograph collection. This also includes application of image indexing retrieval techniques to the CAD system.

Chapter 3 depicts the work on the second HCI-CBR system, using a local model network (LMN). The LMN implements a mixture of Gaussian models through relevance feedback (RF) learning algorithms. It discusses the theoretical principles, and introduce a new learning strategy to apply this network model to interactive retrieval applications.

Chapter 4 presents the work on the machine-controlled content-based retrieval (MCI-CBR) system. The self-organizing system is introduced by incorporating the Self-Organizing Tree Map (SOTM) [35, 36] to the HCI-CBR methods discussed in Chapter 2 and 3. These automatic and semiautomatic systems are applied to web-based databases with the use of compressed-domain descriptors such as discrete cosine transform (DCT) and discrete wavelet transform (DWT) domain visual descriptors.

Chapter 5 presents a new architecture of MCI-CBR on peer-to-peer network, providing for network resource allocations. The chapter also discusses the use of knowledge in MCI-CBR and applies it to a photograph collection.

Chapter 6 describes the work on video indexing and retrieval. First, it introduces adaptive video indexing (AVI) technique to characterize the video database of a news program. Then the chapter presents the integration of AVI to a self-training neural network [43] to implement automatic relevance feedback.

Chapter 7 firstly discusses the application of AVI to an online system for movie retrieval using automatic and semiautomatic relevance feedback. The chapter then describes the audio-visual fusion model to support concept-based queries from movie database.

Chapter 2

INTERACTIVE CONTENT-BASED IMAGE RETRIEVAL

1. Introduction

The central problems regarding the retrieval task are concerned with "interpreting" the contents of the images in a collection and ranking these according to the degree of relevance to the user query. This 'interpretation' of image content involves extracting content information from the image and using this information to match the user's needs. Knowing how to extract this information is not the only difficulty; another is knowing how to use it to decide *relevance.* The decision of relevance characterizing *user information need* is a complex problem.

To be effective in satisfying user information need, a retrieval system must view the retrieval problem as 'human-centered', rather than 'computer-centered'. In a number of recent papers [3–6],an alternative to the computer-centered predicate was proposed. This new approach is based on a human-computer interface which enhances the system to perform retrieval tasks in line with human capabilities. The main activities in this approach consist in analyzing a user's goals from feedback information on the desired images, and adjusting the search strategy accordingly. Here, the user manages the retrieval system, via the interface, through the selections of information gathered during each interactive session, to address information needs which are not satisfied by a single retrieved set of images.

The human-computer interface has been less understood than other aspects of image retrieval, partly because humans are more complex than computer systems, since motivations and behaviors are more difficult to measure and characterize. Recently, studies have been conducted to simulate human perception of visual contents via the use of the supervised analysis method. "Themes" are derived from similarity functions through the assignment of numerical *weights* to the pre-extracted features. The weighted Euclidean is typically adopted to characterize the differences between images, so that distinct weights have varying relevance when used in the simulations (see [6–8] for examples). This idea can be further generalized by incorporating limited adaptivity in the form of a relevance feedback scheme [3–5, 9–11]. Here, weighting is modified according to the user's preference. However, from the user's viewpoint, the limited degree of adaptivity offered, and the restriction of the distance measure to a quadratic form, is not adequate for modeling perceptual difference.

Application of Nonlinear Human-Controled Interactive (HCI) Retrieval

In this chapter, a *nonlinear* approach is presented to address some of these problems to simulate human perception in human-controlled interactive CBR (HCI-CBR). This effectively bridges the gap between the low-level features used in retrieval and the high-level semantics in human perception. The traditional relevance feedback is replaced by a specialized radial-basis function (RBF) network [12, 13] for learning the user's notion of similarity between images. In each interactive retrieval session, the user is asked to separate, from the set of retrieved images, those which are more similar to the query image from those which are less similar. The feature vectors extracted from these classified images are then used as training examples to determine the centers and widths of the different RBFs in the network. This concept is adaptively re-defined in accordance with different user's preferences and different types of images, instead of relying on any pre-conceived notion of similarity through the enforcement of a fixed metric. Compared to the conventional quadratic measure, and the limited adaptivity allowed by its weighted form, the current approach offers an expanded set of adjustable parameters, in the form of RBF centers and widths. This allows a more accurate modeling of the notion of similarity from the user's viewpoint.

The new HCI-CBR is applied to two image database applications. The first domain application is texture retrieval. In this domain, content-based image retrieval is very useful for effective querying using texture patterns that represent a region of interest in a large collection of satellite air photos, as demonstrated in [24, 29, 30]. In the second domain, the HCI-CBR is integrated with an interactive search engine, the Interactive-based Analysis and Retrieval of Multimedia system (iARM) [144], to support image searching tools in large image collections over the internet.

In this chapter, the content-based image retrieval (CBIR) techniques also applied to a computer aided referral (CAR) system. This system implements CBIR process to assist operators for the detection of underwater mine-like objects in side-scan sonar images.

2. Interactive Framework

The most important part in the interactive process is to analyze the role of the user in perceiving image similarity according to preferred image selections. To perform this analysis, a nonlinear model is employed to establish the link between human perception and distance calculation.

In general, learning systems implement a mapping $f_s : \mathcal{R}^P \to \mathcal{R}$ which is given by:

$$y_s = f_s(\mathbf{x}) \tag{2.1}$$

where $\mathbf{x} = [x_1, ..., x_P]^T \in \mathcal{R}^P$ is the input vector corresponding to an image in the database. The main procedure is to obtain the mapping function f_s from a *small* set of training images, $\{(\mathbf{x}_1, l_1), (\mathbf{x}_2, l_2), ..., (\mathbf{x}_N, l_N)\}$, where the two-class label l_i can be in binary or non-binary form.

As many attempts to perform similarity analyzing have focused on linear models, an introduction to linear-based approaches and their limitations are in order. These can be organized into two categories: (1) an approach based on a query reformulation model ([4, 9, 11]); and (2) an approach based on an adaptive metric model ([7, 8, 14, 70]).

Query Reformulation Method

Among the early attempts in the interactive CBIR systems, MARS-1 (Multimedia Analysis and Retrieval System, version 1) [5, 9] implemented the mapping in the form of the query reformulation model (originally proposed by Salton [25] for text retrieval),

$$y_s = f_{\text{cosine}}(\mathbf{x}, \mathbf{x}_{\hat{q}}) \tag{2.2}$$

$$\mathbf{x}_{\hat{q}} = \alpha \mathbf{x}_q + \beta \left(\underset{l_i=1}{mean}\{\mathbf{x}_i\} \right) - \varepsilon \left(\underset{l_i=0}{mean}\{\mathbf{x}_i\} \right) \tag{2.3}$$

where f_{cosine} denotes the cosine measure; $\mathbf{x}_q$ denotes the original query vector; $\mathbf{x}_{\hat{q}}$ denotes the modified query vector; and $(\alpha, \beta, \varepsilon)$ are suitable parameters. The query model $\mathbf{x}_{\hat{q}}$ is obtained by adjusting the positive and negative weight *terms* of the original query $\mathbf{x}_q$. Although simple, this model has been widely used for adaptive information retrieval (IR) [25, 27] and many image retrieval systems [4, 11]. A chief disadvantage of this integration model is the requirement of an indexing structure to follow term-weighting models used in text retrievals for greater effectiveness. Specifically, the model works on the assumption that the query index terms are sparse and are usually of a binary vector representation. However, in image content-based retrieval, vectors are mostly real vectors. In order to overcome this problem, Muller *et al* [11] utilize more than 80,000 feature variables to characterize each image. This method, however, increases computational complexity.

Adaptive Similarity Function

In its later form, weight distance is a common strategy for obtaining the mapping function. This is the case in [6, 10, 64–67, 70, 75], and in the MARS-2 (Multimedia Analysis and Retrieval, version 2) system [8]. In general, the similarity function may be described as:

$$f_s(\mathbf{x}, \mathbf{x}_q) = \sum_{i=1}^{P} h(d_i) \tag{2.4}$$

$$= (\mathbf{x} - \mathbf{x}_q)^T W (\mathbf{x} - \mathbf{x}_q) \tag{2.5}$$

where $h(d_i)$ denotes a one-dimensional transfer function of distance $d_i = |x_i - x_{qi}|$, and W is a *block-diagonal* matrix with the following structure:

$$W = diag[w_1, w_2, ..., w_P] \tag{2.6}$$

The weight parameters $w_i, i = 1, ..., P$ are called *relevance weights*, and are applied to the distance d_i, with the restriction $w_i > 0$, $\sum_i w_i = 1$. These can be estimated by the standard deviation criterion as in [8, 65, 66] or a probabilistic feature relevance method [6].

In addition, different types of basis function have been used [68–70]. Sclaroff *et al* [70] introduced an algorithm for selecting appropriate Minkowski distance metrics according to a minimum distance within the relevant class. This is based on the assumption that metrics are optimal for different classes of query images. With a similar assumption, Bhanu *et al* [67] developed the algorithm for selecting appropriate metrics using reinforcement learning. Choi *et al* [69] proposed an improved similarity function by taking into account the interdependencies between feature elements. This method applies a fuzzy measure over the set of features extracted from feedback samples, and uses the Choquet Integral as the similarity function.

Query and metric Adaptations

In [14, 71–74], an adaptive system combines a query reformulation model with the similarity function (e.g. Eqs.(2.3)-(2.4)) in order to speed up convergence. Some of these works also implemented a query shifting model, using different techniques, such as linear discrimination analysis [74], and a probabilistic distribution analysis on the positive and negative samples [72].

Rui *et al* [14] have derived an optimum solution for the similarity function (Eq.(2.4)), and the query shifting model. This method is referred to as an optimal learning relevance feedback (OPT-RF) method. Using Lagrange multipliers, an optimum solution for a query model is the weighted average of the training samples:

$$\mathbf{x}_{\hat{q}} = \frac{\mathbf{X}^T \mathbf{v}}{\sum_{i=1}^{N} v_i} \tag{2.7}$$

where $\mathbf{v} = [v_1, v_2, ..., v_N]$ are the similarity scores specified by the user, and $\mathbf{X}$ denotes an $N \times P$ matrix, $\mathbf{X} = [\mathbf{x}_1 ... \mathbf{x}_N]^T$. The optimum solution for the weight matrix W is obtained by:

$$W = \begin{cases} (\det(C))^{\frac{1}{K}} C^{-1} & \det(C) \neq 0 \\ diag(\frac{1}{C_{11}}, \frac{1}{C_{22}}, ..., \frac{1}{C_{PP}}) & \text{otherwise} \end{cases} \tag{2.8}$$

where C denotes the weighted covariance matrix,

$$C_{rs} = \frac{\sum_{i=1}^{N} v_i (x_{ir} - x_{\hat{q}r})(x_{is} - x_{\hat{q}s})}{\sum_{i=1}^{N} v_i}, \quad r, s = 1, ..., P \tag{2.9}$$

OPT-RF intelligently switches W between a full matrix and a diagonal matrix, to overcome possible singularity issues when the number of training samples, N, is smaller than the dimensionality of the feature space, P. However, this situation does not usually happen in image retrieval—particularly when images are modeled by multiple descriptors and when only a *small* size of training samples is preferred, i.e., $N < P$.

Ashwin *et al* [71] implement an improved OPT-RF method. Here, the negative samples are used to modify the weight parameters computed, as in Eqs.(2.8)-(2.9), so that the ellipsoids represented by the similarity metric (Eq.(2.5)) better capture more positive samples while excluding negative ones.

Nonlinear Model-based Relevance Feedback

The methods outlined above are referred to as linear-based learning that restricts the mapping function to quadratic form, which cannot cope with a complex decision boundary. Although these learning methods provide a mathematical framework for evaluating image similarity, they are not adequate for the nonlinear nature of human perception. For instance, one-dimension distance mapping $h(d_i)$ in Eq.(2.4) takes the following form:

$$h(d_i) = w_i d_i^2 \tag{2.10}$$

It has no nonlinear capacity, such as:

$$\frac{\partial f_s(\mathbf{x})}{\partial d_i} = 2 w_i d_i \tag{2.11}$$

where w_i is "fixed" to a numerical constant. That is, the linear mapping shows that the degree of similarity between two images is linearly proportional to the magnitudes of their distances. In comparison, the assumption for nonlinear approach is that the same portions of the distances do not always give the same degree of similarity when judged by humans [17].

The visual section of the human brain uses a nonlinear processing system for tasks such as pattern recognition and classification [12]. The application of nonlinear criterion is therefore suitable in performing a simulation task. The current work is different from MARS-2 and OPT-RF in two aspects. First, a nonlinear kernel is applied for the evaluation of image similarity. Second, both positive and negative feedback strategies are utilized for greater effectiveness of the learning capability. By embedding these two properties, the current retrieval system shows a high performance in learning with small user feedback samples, and convergence occurs quickly (results shown in Section 5-6).

The Basic Model

To simulate human perception, a radial basis function (RBF) network [12, 13] is employed as a nonlinear model for proximity evaluation between images. The nonlinear model is constructed by an input-output mapping function, $f(\mathbf{x})$, that uses feature values of input image $\mathcal{X}$ to evaluate the degree of similarity (according to a given query), by a combination of activation functions associated as a nonlinear transformation.

The input-output mapping function, $f(\mathbf{x})$, is performed on the basis of a method called *regularization* [19]. In the context of a mapping problem, the idea of regularization is based on the *a priori* assumption about the form of the solution (i.e., the input-output mapping function $f(\mathbf{x})$). In its most common form, the input-output mapping function is *smooth*, in the sense that similar inputs correspond to similar outputs. In particular, the solution function that satisfies this regularization problem is given by the expansion of the radial basis function [20]. Based on the regularization method, a one-dimensional Gaussian-shaped radial basis function is utilized to form a basic model:

$$G(x) = \exp\left(-\frac{(x-z)^2}{2\sigma^2}\right) \tag{2.12}$$

where z denotes the center of the function and σ denotes its width. The activity of function $G(x)$ is to perform a Gaussian transformation of the distance $d = |x - z|$, which describes the degree of similarity between the input x and center of the function.

To estimate the input-output mapping function $f(\mathbf{x})$ the Guassian RBF is expanded through both its center and width, yielding different RBFs which are then formed as an RBF network. Its expansion is implemented via interactive learning, where the expanded RBFs can optimize weighting, to capture human perception similarity as discussed in the following section.

3. Radial Basis Function (RBF)-Based Relevance Feedback (RF)

Radial basis function (RBF) networks possess an excellent nonlinear approximation capability [12, 13]. This property is utilized to design a system of locally tuned processing units to approximate the target nonlinear function $f(\mathbf{x})$. In the general solution, an approximation function obtained by the RBF networks takes the following form:

$$f(\mathbf{x}) = \sum_{j=1}^{N} w_j G(\mathbf{x}, \mathbf{z}_j) \tag{2.13}$$

$$= \sum_{j=1}^{N} w_j \exp\left(-\frac{1}{2\sigma_j^2}\sum_{i=1}^{P}(x_i - z_{ji})^2\right) \tag{2.14}$$

where $\mathbf{z}_j \in \mathcal{R}^P$ denotes the center of the function $G(\mathbf{x}, \mathbf{z}_j)$, σ_j denotes its width, and $\mathbf{x} \in \mathcal{R}^P$ denotes the input vector. There are N Gaussian units in this network. Their sums, in the form of a linear superposition, define the approximating function $f(\mathbf{x})$. With the *regularization* structure, the RBF network takes a one-to-one correspondance between the training input samples and the function $G(\mathbf{x}, \mathbf{z}_j)$, by which each training sample is associated with the center $\mathbf{z}_j, j \in \{1, N\}$.

A direct application of this network structure to online learning image retrieval is, however, considered prohibitively expensive to implement in computational terms for large N. It is also sufficient to reduce the network structure into a single unit, since image relevance identification requires only a two-class separation for a given query.

In the current work, with radial-basis functions in mind, a one-dimensional Gaussian-shaped RBF is associated with each component of the feature vector, as follows:

$$f(\mathbf{x}) = \sum_{i=1}^{P} G_i(x_i, z_i) \tag{2.15}$$

$$= \sum_{i=1}^{P} \exp\left(-\frac{(x_i - z_i)^2}{2\sigma_i^2}\right) \tag{2.16}$$

where $\mathbf{z} = [z_1, ..., z_i, ..., z_P]^T$ is the adjustable query position or the center of the RBF function, $\sigma_i, i = 1, ..., P$ are the tuning parameters in the form of RBF widths, and $\mathbf{x} = [x_1, ..., x_i, ..., x_P]^T$ is the feature vector associated with an image in the database. Each RBF unit implements a Gaussian transformation which constructs a local approximation to a nonlinear input-output mapping. The magnitude of $f(\mathbf{x})$ represents the similarity between the input vector $\mathbf{x}$ and the query $\mathbf{z}$, where the highest similarity is attained when $\mathbf{x} = \mathbf{z}$. Based on simulation results in Section 5, the new single unit RBF network is effective in learning and quickly converges for one-class-relevance classification using small volume of training sets.

Expansion of RBFs

In the network structure, each RBF function is characterized by two adjustable parameters, the tuning parameter and the adjustable center:

$$\{\sigma_i, z_i\}, i = 1, ..., P, \tag{2.17}$$

to form a set of P basis functions,

$$\{G_1(\sigma_1; z_1), G_2(\sigma_2; z_2), ..., G_P(\sigma_P; z_P)\}. \tag{2.18}$$

These parameters are estimated and updated via learning algorithms. The first assumption behind the learning algorithms, is that the user's judgment of image similarity can be captured by a small number of pictorial features. This is an unequal bias toward the evaluation of image similarity. That is, given a semantic context, some pictorial features exhibit greater importance or "relevance" than others in the proximity evaluation. This is the same assumption which underlies image matching algorithms in [21, 6]. However, in this case, the weighting process is controlled by an expanded set of tuning parameters, $\sigma_i, i = 1, ..., P$, which reflects the relevance of individual features. If a feature is highly relevant, the value of σ_i should be small to allow greater sensitivity to any change of the distance $d_i = |x_i - z_i|$. In contrast, a large value of σ_i is assigned to the non-relevant features so that the corresponding vector component can be disregarded when determining its similarity, since the magnitude of $G_i(\cdot)$ is approximately equal to unity regardless of the distance d_i. The choice of σ according to this criterion will be discussed in Section 4.0.

The second assumption concerns the relationship between the clustering of desired images in the P-dimensional feature space and the initial location of the query. For a given query image, the associated feature vector may not be in a position close enough to those stored vectors associated with other relevant images. This initial query may form a decision region that contains only a local cluster of the desired images in the database. The goal here, then, is to associate this local cluster as prior information in order to describe a larger cluster of relevant images in the database. The description of this larger cluster of relevant images is built interactively with assistance from the user. This process is implemented by the RBF network through the adjustment of RBF centers, $z_i, i = 1, ..., P$, as will be described in the following section.

4. Learning and Characterization

The learning algorithms enable the RBF network to progressively model the notion of image similarity for effective searching. The image matching process is initiated when the user supplies a query image and the system retrieves the N_c images in the databases which are closest to the query image. From these images the user selects those as relevant which are most similar to the current query image, while the rest are regarded as nonrelevant. The feature vectors extracted from these images are incorporated as training data for the RBF network in order to modify the centers and widths. The re-estimated RBF model is then used to evaluate the perceptual similarity in a new search, and the above process is repeated until the user is satisfied with the retrieval results.

Center selection

Given a set of images, the human user may easily distinguish the relevant and nonrelevant images according to their own information needs (Fig. 2.1(a)). In contrast, a computer interprets relevance as the distance between low-level image features (Fig. 2.1(b)), which could be very different from that shown in Fig. 2.1(a). The low-level vector of the query is likely to be located in a different position in the feature space and may not be a representative sample of the relevant class. To improve computer retrieval performance, the low-level query vector is modified via the learning algorithm. This aims at optimizing the current search. The expected effect is that the new query will move towards the relevant items (corresponding to the desired images) and away from the non-relevant ones, whereas the user's information need remains the same throughout the query modifying process. In the following discussion, the basic optimization procedure known as learning vector quantization (LVQ) [12] is firstly described. Then, a modified LVQ is presented to obtain a proper choice for the RBF center associated with the new query vector.

LVQ

LVQ [12] is a supervised learning technique used to optimize vector structures in a *code book* for the purpose of data compression [23]. The initial vectors (in a codebook), referred to as Voronoi vectors, are modified in such a way that all points partitioned in the same Voronoi cells have the minimum (overall) encoding distortion. The technique uses the class information provided in a training set to move the Voronoi vectors slightly, so as to improve the accuracy of classification. Let the input vector $\mathbf{x}$ be one of the samples in the training set. If the class labels of the input vector $\mathbf{x}$ and a Voronoi vector $\mathbf{z}$ agree, the Voronoi vector $\mathbf{z}$ is moved in the direction of the input vector $\mathbf{x}$. On the other hand, if the class labels of the input vector $\mathbf{x}$ and the Voronoi vector $\mathbf{z}$ disagree, the Voronoi vector $\mathbf{z}$ is moved away from the input vector $\mathbf{x}$.

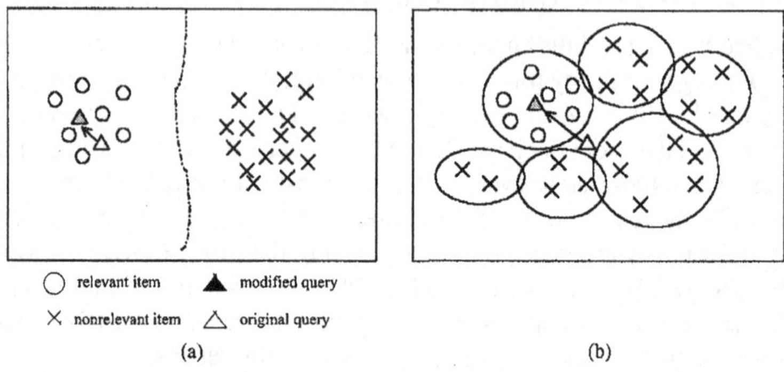

Figure 2.1. Query modification; (a) relevance judgment based on human vision; (b) relevance clustering in the feature space. In (a), given an image collection set, the human user may easily distinguish the relevant images from the high-level semantics according to his/her own understanding and expression of require information. In contrast, the low-level feature vector of the query in (b) is likely to be located in a different position in the feature space and may not be a representative sample of the relevant class.

The modification of the Voronoi vectors is usually carried out by an iterative process, where $n = 0, 1, 2, ..., n_{\max-1}$ is the step index. Let $\{\mathbf{z}_j\}_{j=1}^{J}$ denote the set of Voronoi vectors. Also, let $\{\mathbf{x}_i\}_{i=1}^{N}$ denote the set of training samples. First, for each input vector $\mathbf{x}_i(n)$, the index $c(\mathbf{x}_i)$ of the best-matching Voronoi vector $\mathbf{z}_c(n)$ is identified by the condition:

$$c = \arg\min_j \{||\mathbf{x}_i - \mathbf{z}_j||\} \tag{2.19}$$

Let $\ell_{\mathbf{z}_c}$ denote the class label associated with the Voronoi vector $\mathbf{z}_c$, and $\ell_{\mathbf{x}_i}$ denote the class label of the input vector $\mathbf{x}_i$. The Voroinoi vector $\mathbf{z}_c$ is adjusted as follows:

If $\ell_{\mathbf{z}_c} = \ell_{\mathbf{x}_i}$, then (*reinforced learning*)

$$\mathbf{z}_c(n+1) = \mathbf{z}_c(n) + \alpha_n[\mathbf{x}_i(n) - \mathbf{z}_c(n)] \tag{2.20}$$

If, on the other hand, $\ell_{\mathbf{z}_c} \neq \ell_{\mathbf{x}_i}$, then (*antireinforced learning*)

$$\mathbf{z}_c(n+1) = \mathbf{z}_c(n) - \alpha_n[\mathbf{x}_i(n) - \mathbf{z}_c(n)] \tag{2.21}$$

Note that, for all $j \neq c$, $\mathbf{z}_j(n+1) = \mathbf{z}_j(n)$, those Voronoi vectors remain unchanged. Here, the learning constant, α_n, decreases monotonically with the number of iterations and where $0 < \alpha_n < 1$. After several passes through the training data, the Voronoi vectors typically converge, and the training process is completed.

Based on the reinforced learning rule, it is clearly shown that the above process tries to move the Voronoi vector $\mathbf{z}_c$ to points in the input space that are close to those samples which have the same class labels. At the same time, the antireinforced learning rule moves $\mathbf{z}_c$ away from those samples which are in different classes. This process results in a new set of Voronoi vectors, $\{\tilde{\mathbf{z}}_j\}_{j=1}^{J}$, that minimizes (overall) encoding distortion.

A Modified LVQ

In an interactive retrieval session, it is desirable to a reduce the processing time to minimum without affecting the overall performance. So, the LVQ method can be considered for query modification, without the implementation of the *iterative* procedure. This minimizes the time complexity of the process $\mathcal{O}(n_{max})$, where n_{max} is the total number of iterations.

In image retrieval, the database feature space can be clustered into a number of distinct Voronoi cells with associated Voronoi vectors. Furthermore, the Voronoi vectors may be individually initialized by query vectors. Each Voronoi cell contains a set of feature vectors associated with those retrieved images that are the closest to the corresponding query, according to the nearest-neighbor rule based on the Euclidean metric. The objective, here, is to optimize these cells by employing the LVQ algorithm. Since only one query is submitted at a particular time, only two partitions are necessary in the space, with one representing the relevant image set. The LVQ algorithm is then adopted to modify this cell from its corresponding training data.

Let the Voronoi vector $\mathbf{z}_q(t)$ denote the submitted query at a retrieval session t. Recall that the information input from the user at the interactive cycle is formed as the training set $\mathcal{T}$ that contains training vectors belonging to two separate classes:

$$\begin{aligned} \mathcal{T}_{(t)} &= (\mathbf{x}_i, l_i),\ i = 1, \ldots, N && (2.22)\\ &= \{\mathbf{x}'_m, \mid l_m = 1\} \cup \{\mathbf{x}''_k, \mid l_k = 0\} && (2.23)\\ m &= 1, ..., M && (2.24)\\ k &= 1, ..., K && (2.25) \end{aligned}$$

where $\mathbf{x}_i \in \mathcal{R}^P$ is a feature vector; $l_i \in \{0, 1\}$ is a class label; $\mathbf{x}'$ and $\mathbf{x}''$ are the positive and negative sample, respectively. The set of vectors in Eq.(2.22) represents the set of points closest to the submitted query, $\mathbf{z}_q(t)$, according to the distance calculation in the previous search operation. Consequently, each data point can be regarded as the vector $\mathbf{x}_i$ that is 'closest' to the Voronoi vector $\mathbf{z}_q(t)$. Therefore, following the LVQ algorithm, it is observed that all points in this training set are used to modify only the best-matching Voronoi vector, that is, $\mathbf{z}_q(t)$.

Model 1: According to the previous discussion, after the training process is completed the *modified* Voronoi vector $\tilde{\mathbf{z}}$ will lie close to the data points that are in the same class, and away from those points that are in a different class. Combining these ideas, a modified LVQ algorithm is now obtained, to adjust the query vector $\mathbf{z}_q(t)$, by approximating the *modified* Voronoi vector $\tilde{\mathbf{z}}_q$ upon convergence:

$$\mathbf{z}_q(t+1) = \mathbf{z}_q(t) + \alpha_R(\overline{\mathbf{x}}' - \mathbf{z}_q(t)) - \alpha_N(\overline{\mathbf{x}}'' - \mathbf{z}_q(t)) \quad (2.26)$$

$$\overline{\mathbf{x}}' = \frac{1}{M}\sum_{m=1}^{M} \mathbf{x}'_m \quad (2.27)$$

$$\overline{\mathbf{x}}'' = \frac{1}{K}\sum_{k=1}^{K} \mathbf{x}''_k \quad (2.28)$$

Where $\mathbf{z}_q(t)$ is the previous query, $\mathbf{x}'_m = [x'_{m1}, \ldots, x'_{mi}, \ldots, x'_{mP}]^T$ is the m-th feature vector of relevant images; $\mathbf{x}''_k = [x''_{k1}, \ldots, x''_{ki}, \ldots, x''_{kP}]^T$ is the k-th feature vector of nonrelevant images; α_R and α_N are suitable positive constants; and M and K are, respectively, the number of relevant and nonrelevant images in the training set. The application of the query modification in Eq.(2.26) is to allow the new query, $\mathbf{z}_q(t+1)$, to move towards the new region populated by the relevant images as well as to move away from those regions populated by non-relevant images.

Eq.(2.26) is illustrated in Fig. 2.2. Let the centers of the relevant image set and nonrelevant image set in the training data, be R and N, respectively. Also, let $z_q(t) = z_c$. As shown in Fig. 2.2, the effect of the second term on the right hand side of Eq.(2.26) is to allow the new query to move towards R. If in $N = N_1 < z_q(t)$, the third term is negative; so, the current query will move to the right, i.e., the position of $z_q(t)$ will shift away from N_1 to $\hat{z}_1$. On the other hand, when $N = N_2 > z_q(t)$, the third term is positive, hence $z_q(t)$ will move to the left or $\hat{z}_2$; i.e., away from N_2 .

In practice one finds that the relevant image set is more important in determining the modified query than the nonrelevant images. The is because the set of relevant images is usually tightly clustered due to the similarities among its member images, and thus satisfies the modified query with little ambiguity. This is illustrated in Fig 2.3(a). On the other hand, the set of nonrelevant images is much more heterogeneous, therefore, the centroid of this nonrelevant image set may be located almost anywhere in the feature space. As a result, $\alpha_R > \alpha_N$ is chosen for Eq.(2.26) to allow a more definite movement toward the set of relevant images, while permitting slight movement away from the non-relevant regions.

The current approach works well when the sets of relevant and non-relevant images are well-separated, as in Fig. 2.3(a). In practice, the set of non-relevant

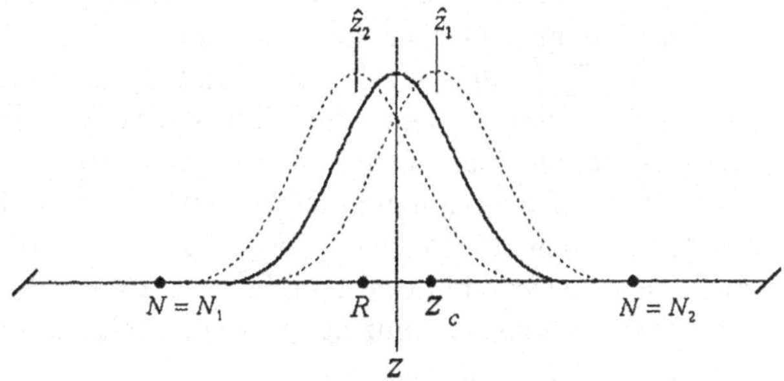

Figure 2.2. Illustration of query modification.

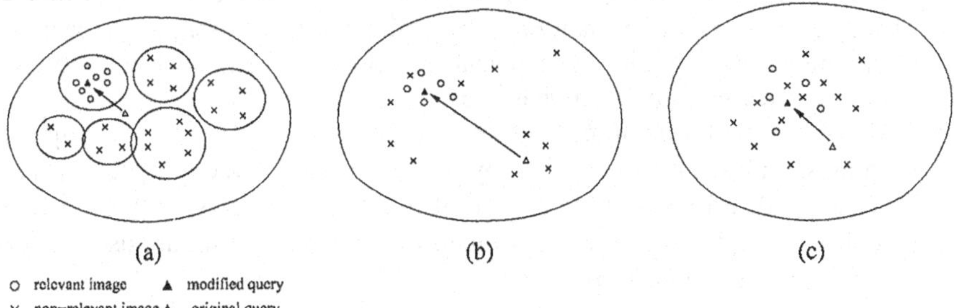

Figure 2.3. Query modification in the feature space; (a) ideal configuration; (b) favorable configuration; (c) unfavorable configuration.

images usually covers a wider region of the space, as shown in Figs. 2.3(b) and (c). The effectiveness of the current approach will thus depend on the exact distribution of the non-relevant images in the space. Fig. 2.3(b) illustrates a particular distribution which is favorable to the current approach, while the case illustrated in Fig. 2.3(c) may compromise performance.

Model 2: In order to provide a simpler procedure and a direct movement of the new query towards the relevant set, Eq.(2.26) is reduced to

$$\mathbf{z}_q(t+1) = \overline{\mathbf{x}}' - \alpha_N(\overline{\mathbf{x}}'' - \mathbf{z}_q(t)) \quad (2.29)$$

The first and the second terms in the right hand side of Eq.(2.26) are replaced by $\overline{\mathbf{x}}'$ (centroid of the relevant vectors). Since the relevant image group indicates the user's preference, the presentation of $\overline{\mathbf{x}}'$ for the new query will give a reasonable representation of the desired image. In particular, the mean value,

$\overline{x}'_i = (1/M)\sum_{m=1}^{M} x'_{mi}$, is a statistical measure providing a good representation of the i-th feature component since this is the value which minimizes the average distance $(1/M)\sum_{m=1}^{M}(x'_{mi} - \overline{x}'_i)^2$. Further, the exclusion of the parameter α_R from Eq.(2.26) permits greater flexibility, since only one procedural parameter is necessary for the final fine tuning of a new query.

Finally, a comparison of the optimum query model, based on Eq.(2.7) to Eq.(2.26) and Eq.(2.29), shows the following: Eq.(2.7) is optimal on a criterion of minimum distance to relevant retrievals, while Eq.(2.26) and Eq.(2.29) are multiple criterion optimizations, minimizing distance to relevant samples and maximizing distance to nonrelevant samples.

Selection of RBF width

As observed from the previous discussions, the nonlinear transformation associated with the output unit(s) of the Gaussian-shaped RBF are adjusted in accordance with different user's preferences and different types of images. Through the proximity evaluation, differential biases are assigned to each feature, while features with higher relevance degrees are emphasized, and those with lower degrees are de-emphasized.

Consider that for a particular query location $\mathbf{z} = [z_1, \ldots, z_i, \ldots, z_P]^T$, the training samples can be described by the set of feature vectors $\{\mathbf{x}_i\}_{i=1}^{N}$, as in Eq.(2.22). To estimate the relevance of individual features, only the vectors associated with the set of relevant images in this training set are used to form an $M \times P$ feature matrix $\mathbf{R}$:

$$\begin{aligned}\mathbf{R} &= [\mathbf{x}'_1, \ldots, \mathbf{x}'_m, \ldots, \mathbf{x}'_M]^T \\ &= [x'_{mi}] \quad m = 1, \ldots, M,\ i = 1, \ldots, P\end{aligned} \tag{2.30}$$

where $\mathbf{x}'_m = [x'_{m1}, \ldots, x'_{mi}, \ldots, x'_{mP}]^T$ corresponds to one of the images marked as relevant; x'_{mi} is the i-th component of the feature vector $\mathbf{x}'_m$; P is the total number of features; and M is the number of relevant images. According to previous discussion, the tuning parameter σ_i should reflect the relevance of individual features. It was proposed, in [6, 21], that given a particular numerical value z_i for a component of the query vector, the length of the interval which completely encloses z_i and a pre-determined number L of the set of values x'_{mi} in the relevant set which falls into its vicinity, is a good indication of the relevancy of the feature. In other words, the relevancy of the i-th feature is related to the density of x'_{mi} around z_i, which is inversely proportional to the length of the interval. A large density usually indicates high relevancy for a particular feature, while a low density implies that the corresponding feature is not critical to the similarity characterization. Setting $L = M$, the set of tuning parameters is thus estimated as follows.

$$\sigma_i = \eta \max_m |x'_{mi} - z_i| \tag{2.31}$$

The factor η guarantees a reasonably large output $f(\mathbf{x})$ for the Gaussian RBF unit, which indicates the degree of similarity, e.g., η=3.

The second criterion is also considered for estimating the tuning parameters. This is obtained by nonlinear weighting of the sample variance in the relevant set as follows:

$$\sigma_i = \exp(\beta \cdot \mathrm{STD}_i) \tag{2.32}$$

$$\mathrm{STD}_i = \left(\frac{1}{M-1} \sum_{m=1}^{M} (x'_{mi} - \bar{x}'_i)^2 \right)^{\frac{1}{2}} \tag{2.33}$$

where STD_i is the standard deviation of the members $x'_{mi}, m = 1, \ldots, M$, which is inversely proportional to their density (Gaussian distribution). The parameter β can be chosen to maximize or minimize the influence of STD_i on σ_i. For example, when β is large, a change in STD_i will be exponentially reflected in σ_i.

As a result, Eqs.(2.31)-(2.32) provide a small value of σ_i if the i-th feature is highly relevant, (i.e., the sample variance in the relevant set $\{x'_{mi}\}_{m=1}^{M}$, is small). This allows higher sensitivity to any change of the distance $d_i = |x_i - z_i|$. In contrast, a high value of σ_i is assigned to the non-relevant features, so that the corresponding vector component can be disregarded when determining the similarity.

5. Application to Texture Image Retrieval

These experiments study the retrieval performance of the nonlinear RBF approach to two image retrieval application domains. This section describes the application on texture pattern retrieval, and Section 6 describes the application on a large collection of photographs. When evaluating image-retrieval algorithms, there are several factors that determine the choice of a particular algorithm for an application. Central concerns are retrieval accuracy and CPU time. The retrieval accuracy is evaluated by a specific ground truth on a given database. For the adaptive retrieval algorithms, however, there are additional factors, such as the size of the training set, and the convergence speed. For each domain application, the new RBF algorithm is evaluated and compared to other HCI-CBR systems, using these factors.

Comparison Method

The RBF's retrieval performance was compared to the MARS-1, which was developed early in the texture retrieval domain [9, 119]. The retrieval strategy in MARS-1 has also been extended and used in other works, such as [3, 11]. The applications compared are as follows:

1 The radial basis function (RBF) methods: the RBF1 method uses model 1 for determining the RBF center (Eq.(2.26)), and Eq.(2.32) for the RBF width. The RBF2 method uses model 2 for determining the RBF center (Eq.(2.29)) and Eq.(2.31) for learning RBF width.

2 The relevance feedback method (RFM) is described in the MARS-1 system [9, 119], is employed by the PicToSeek system [3], and is also used in [4, 10, 11].

3 Method 3: simple CBIR uses a non-interactive retrieval method, which corresponds to the first iteration of interactive search. This method employs different types of similarity functions, including weighted distance (as in Eq.(2.35) below), cosine distance, and the histogram intersection, corresponding to the first iteration found in RBF, MARS-1, and PicToSeek.

Databases and Ground Truth Classes

The databases and the corresponding ground truth data are generated in the same ways as in previous works [6, 9, 64, 65, 67], as well as in MPEG-7 Texture and Color Core Experiments [88, 115]. Performance evaluations of retrieval were carried out using two standard texture databases: (1) the MIT texture collections, and (2) the Brodatz database [24]. In the first collection, the original test images were obtained from MIT Media Laboratories [116]. There were 39 texture images from different classes manually established through a process of visual inspection. The large image, 512×512 pixels in size, was divided into 16 non-overlapping sub-images, 128×128 in size, creating a database of 624 texture images (shown in Fig. 2.4). In the second database, texture images and a feature set were created by Ma and Manjunath [24] at the University of California at Santa Barbara (UCSB). The Brodatz database contains 1,856 patterns obtained from 116 different texture classes. Each class contains 16 similar patterns. Fig. 2.4 shows all the texture classes, which include both homogeneous and non-homogeneous textures. The strategy used to obtained this database is also based on tiling a large image into smaller sub-images. These can be observed in Fig. 2.5.

Homogeneous Texture Descriptor (HTD)

The HTD is one of the MPEG-7 texture descriptors described in [88]. HTD characterizes the homogeneous texture pattern based on computing the local spatial-frequency statistics of the texture. Each texture in the database is described by a 48-dimensional HTD vector, which is constructed as follows:

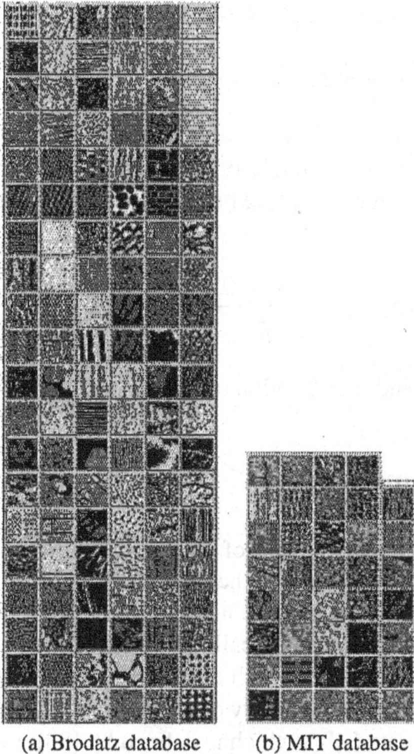

(a) Brodatz database (b) MIT database

Figure 2.4. (a) 116 texture image classes in Brodatz database; (b) 39 texture image classes in MIT database.

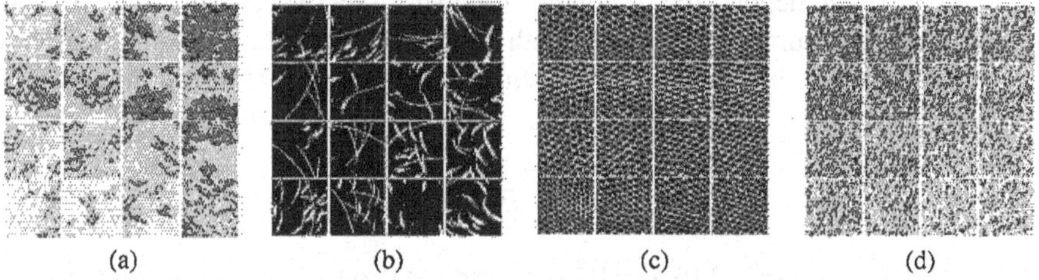

(a) (b) (c) (d)

Figure 2.5. (a)-(b) Some examples of non-homogeneous texture; (c)-(d) some examples of homogeneous texture. In each case, the large image is partitioned into 16 sub-images.

firstly, the Gabor wavelet transform is applied to the image, where the set of basis functions consists of Gabor wavelets spanning four scales and size orientations. The mean and standard deviation of the transform coefficients are then used to form the feature vector. The HTD is described by

$$\mathrm{TD} = [e_1, e_2, \ldots, e_{24}, d_1, d_2, \ldots, d_{24}]^T \tag{2.34}$$

where e_i and d_i represent the mean and standard deviation of the ith feature channel. Since the dynamic range of each feature component is different, a suitable similarity measure for this feature is computed by the following distance measure [24, 88]:

$$d(\mathrm{TD}_{\mathrm{qyery}}, \mathrm{TD}_{\mathrm{Database}}) = \sum_k \left| \frac{\mathrm{TD}_{\mathrm{query}}(k) - \mathrm{TD}_{\mathrm{Database}}(k)}{\alpha(k)} \right| \tag{2.35}$$

where $\alpha(k)$ is the standard deviation of $\mathrm{TD}_{\mathrm{Database}}(k)$ for a given database.

Summary of Comparison

In the simulation study, a total of 39 images, one from each class, were selected as the query images from the MIT database. For each query, the top 16 images were retrieved to provide necessary relevance feedback. Using this method, all the top 16 retrievals ideally are from the same classes. Similarly, a total of 116 images, one from each class, were selected as the query images from the Brodatz database. In the two databases, performance was measured in terms of retrieval rate (RR), which is defined by [88]:

$$\mathrm{RR}(q) = \frac{\mathrm{NF}(a,q)}{\mathrm{NG}(q)} \in [0,1] \tag{2.36}$$

where $\mathrm{NG}(q)$ denotes the size of the ground truth set for a query q; and $\mathrm{NF}(a,q)$ denotes the number of ground truth images found within the first $a = 16$ retrievals. For the whole set of NQ queries, the average retrieval rate (AVR) is given by:

$$\mathrm{AVR} = \frac{1}{\mathrm{NQ}} \sum_{q=1}^{\mathrm{NQ}} \mathrm{RR}(q) \tag{2.37}$$

where $\mathrm{NQ} = 39$ and 116 for MIT database and Brodatze databases, respectively.

Summary of Retrieval on MIT Database

The average retrieval rate of the 39 query images is summarized in Table 2.1, where t denotes the number of iterations. The following observations are based on the results.

Method	t=0	t=1	t=2	t=3	Parameters
RBF1	74.36	90.06	92.95	93.59	$\alpha_R = 1.4, \alpha_N = 0.4, \beta = 2.6$
RBF2	74.36	88.62	91.67	92.79	$\alpha_N = 0.65$
MARS-1	64.26	77.73	79.97	80.13	$(\alpha, \gamma, \varepsilon) = (1, 5, 0.5)$

Table 2.1. Average retrieval rate (%) for the 39 query images in MIT database, using Gabor texture feature representation.

First, for all methods, the performance with interactive learning after 3 iterations (t=3) was substantially better than non-interactive cases ($t = 0$). The improvements are quite striking. *Second*, after 3 rounds of interactive learning, the RBF1 method gave the best performance: on average 93.59% of the correct images are in the top 16 retrieved images (i.e., more than 14 of the 16 correct images are present). This is closely followed by RBF2, at 92.79% of correct retrieval. These results show that the RBF methods perform substantially better than MARS-1, which provides a retrieval performance of 80.13%. It is also observed that the RBF methods provide much better results after one iteration (88.62%) than MARS-1 even after 3 iterations (80.13%). *Third*, for all three interactive methods, convergence is achieved within a few iterations.

Fig. 2.6 shows two retrieval sessions performed by RBF1 in comparison with MARS-1. It clearly illustrates the superiority of the nonlinear method. It was observed that RBF1 considerably enhanced retrieval performance, both visually and statistically. In addition, given the small number of training samples (e.g., 16 retrieved images used in training), the RBF approach can more effectively learn and capture user input on image similarity.

Retrieval on the Brodatz Database

Fig. 2.7(a) summarizes the experimental results obtained from the Brodatz database. It shows the average retrieval accuracy of the different methods for the 116 query images, each of which was selected from a different class. It can be seen that all interactive methods demonstrate significant performance improvement across the task. The final results, after learning, show that RBF1 gave the best performance at 90.5% of correct retrieval, followed by RBF2 (87.6%), with MARS-1 (78.5%) a distant third. It was observed earlier that the characteristics of the retrieval results obtained from the Brodatz database are very similar to those obtained from the MIT database. This implies that RBF1 consistently displays superior performance over MARS-1.

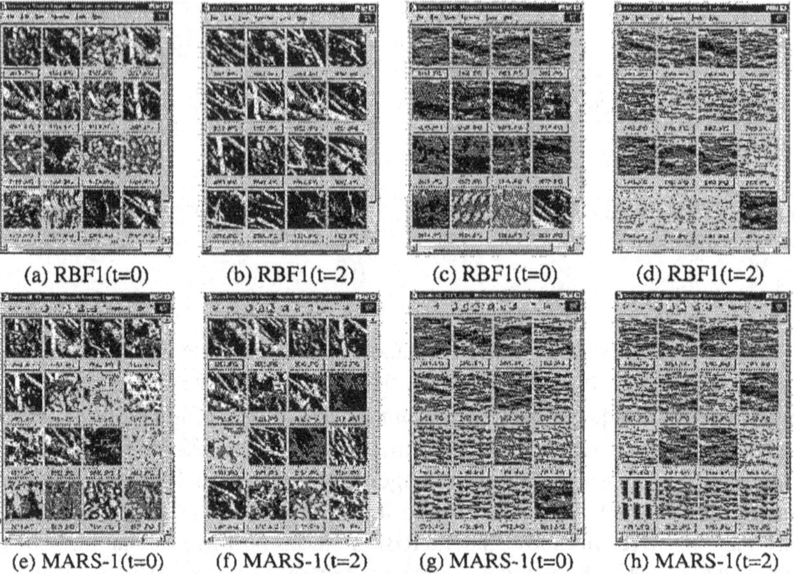

Figure 2.6. Pattern retrieval results before and after learning similarity, using MIT database. Results show a comparison between RBF1 and MARS-1 using Gabor wavelet features; (a), (b), (e), and (f) show retrieval results in the answer to the query 'Bark.0003'; (c), (d), (g), and (h) show retrieval results in the answer to the query 'Water.0000'. In each case, the images are ordered according to decreasing similarity among the 16 best matches, from left to right and top to bottom.

System performance varied greatly depending on the nature of the query. Some queries were easy (i.e., retrieval rate at t=0 is more than 87.5%). Here, the relevant sets were similarly acquired by every method, and the performance after interactive learning was often perfect. By contrast, with a more difficult query, the relevant sets varied greatly in size and composition. In such cases, the effect of interactive learning fluctuated more dramatically within the different methods. To compare the retrieval performances more subjectively, the average retrieval rates were re-calculated excluding 26 query images with a retrieval rate of 100% at t=0. This result is shown in Fig. 2.7(b). Consequently, it was observed that the highest rate was at 87.8% (RBF1), an improvement of 21.7% from 66.1%.

Fig. 2.8 illustrates retrieval examples with and without learning similarity. It shows some of the difficult patterns analyzed, which clearly illustrates the superiority of the RBF method.

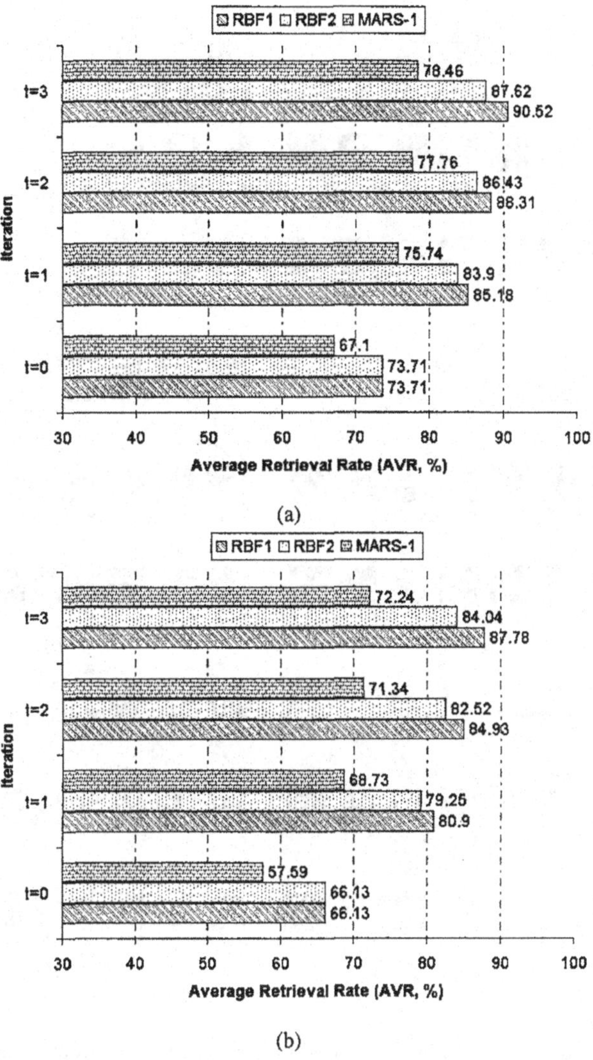

Figure 2.7. Average Retrieval Rate, AVR (%) obtained by retrieving 116 query images in the Brodatz database, using Gabor wavelet representation: (a) AVR of all 116 queries, (b) AVR of (a) excluding 26 query images that had a rate of 100% at t=0.

6. Application to Digital Photograph Collection

In this section, the interactive system is applied to photograph collection from the Corel Gallery 65000 product [22]. The database contains 40000 real-life

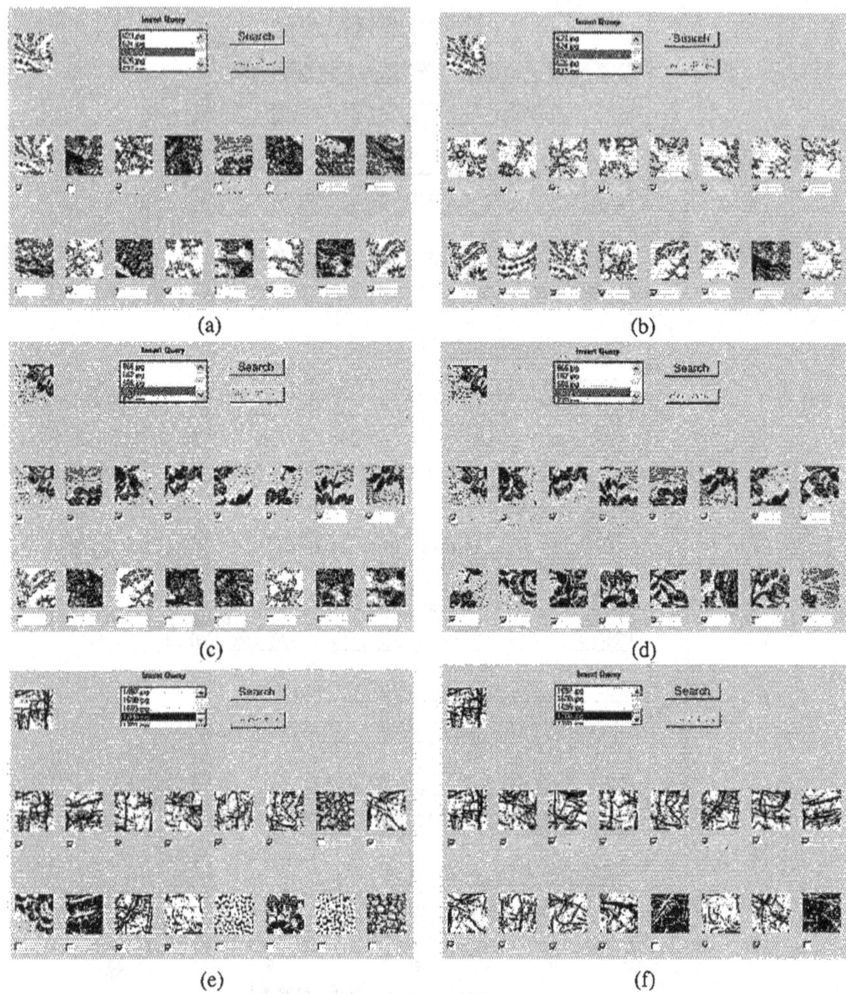

Figure 2.8. Top sixteen retrievals obtained by retrieving textures D625, D669, D1700, and D240 from Brodatz database, using RBF1. Image on the left, (a), (c), and (e) show results before learning, and on the right, (b), (d), and (f), show results after learning.

photographs, in two groups, each of which is either 384 × 256 or 256 × 384 pixels in size (shown in Fig. 2.9). It is organized into 400 categories by Corel professionals. These categories were used as a ground truth in this evaluation. For indexing purposes, each image is characterized by visual descriptors using multiple types of features, $F = \{F_{color}, F_{texture}, F_{shape}\}$ where the representations are color histogram [98] and color moments for color descriptors;

GW transform for texture descriptors [24]; and Fourier descriptor for shape descriptors [31]. The algorithms for obtaining these descriptors are summarized in Table 2.2. Note that the resulting feature database, which is a matrix of size $M \times D$, ($M = 40000$ and $N = 114$), was scaled by feature mean values and standard deviations to remove unequal dynamic ranges of each feature variable.

The following simulations shows performance comparisons between the nonlinear RBF method, MARS-2 [8] and OPT-RF [14] systems (described in Section 2). MARS-2 is relatively newer than MARS-1, and has been intensively tested on the large Corel image collection in [8]. This has become a popular benchmark for image retrieval. In [14], OPT-RF has recently proven to be the optimization framework, more so than in previous studies on interactive CBIR systems. The major difference between these two systems is that the learning algorithm in OPT-RF has both an optimum query and a switching option of the weight matrix W (cf. Eq.(2.8)) between a full matrix and a diagonal matrix. Particularly in this practical application, since the feature dimensions are very high ($D = 114$), OPT-RF was implemented with W in a diagonal matrix form. In the RBF case, relevance feedback learning is processed based on the Gaussian kernel, having a nonlinear decision criterion. In addition, RBF obtains automatic weighting to capture user perception, whereas OPT-RF requires users to specify weighting along a slider bar (cf. Fig. 2.11). Moreover, the RBF method uses both positive and negative samples to track the optimum query model. Neither OPT-RF nor MARS-2 support these features.

The average precision rates (APR) and CPU time required are summarized in Table 2.3. These are obtained using the RBF method (Eq.(2.29) and Eq.(2.32)), MARS-2 system (Eqs.(2.4)-(2.6)), and OPT-RF system (Eqs.(2.7)-(2.8)). Notice that all methods employ norm-1 metric distance to obtain initial retrieval results at $t = 0$. A total of 35 queries were selected from different categories. The performances were measured from the top 16 retrievals, and averaged over all 35 queries.

Evidently, the nonlinear RBF method exhibits significant retrieval effectiveness, while offering more flexibility than MARS-2 and OPT-RF. With this large, heterogeneous image collection, an initial result obtained by the simple CBIR system has less than 50% precision. With the application of RBF interactive learning, the performance can be improved to greater than 90% precision. Due to the limitation in the degrees of adaptivity, MARS-2 provides the lowest performance gains and converges at about 62% precision. It is observed that the learning capability of RBF is more robust than that of OPT-RF, not only in retrieval accuracy, but also learning speed. As presented in Table 2.3, results after *one* round of the RBF is similar to results after *three* rounds of the

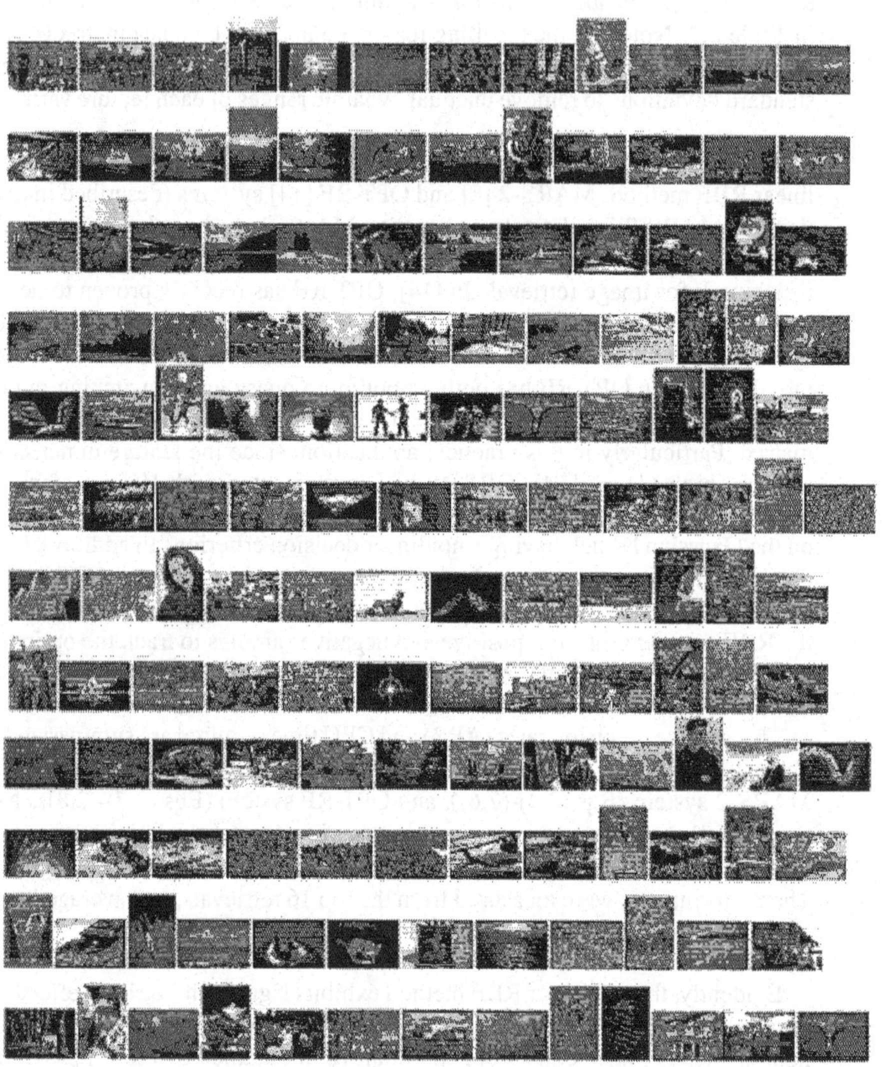

Figure 2.9. Example images from Corel database.

OPT-RF. This quick learning is highly desirable, since the user workload can be minimized. This robustness follows from imposing nonlinear discriminant capability in combination with positive and negative learning strategies. Notice that OPT-RF requires the user to specify weight parameters in the form of a slider bar for learning, whereas RBF automatically evaluates these weight parameters from the fed-back images.

In regard to CPU time for the retrievals, the RBF approach is longer, at 2.34 seconds per iteration for a single query. However, the RBF method gains about 80% precision within only the first iteration, i.e., in only 2.34 seconds. By contrast, though faster, the OPT-RF needs three iterations to reach this underlined performance, i.e., taking $1.27 \times 3 = 3.81$ seconds. In other words, RBF can reach the best performance within a shorter CPU time than the other methods discussed. This also means that OPT-RF users are required to go through two more rounds of feedback in order to achieve equivalent performance. Furthermore, when subject to three iterations, RBF reaches a 91% precision level that cannot be achieved by any other method.

Typical retrieval sessions are shown in Figs. 2.10-2.13. Fig. 2.10 shows retrieval results of the "Yacht" query. Fig. 2.10(a) shows the 16 best-matched images before applying any feedback, with the query image display in the top-left corner. It was observed that some retrieved images are similar to the query in terms of color composition. In this set, three retrieved images were marked as relevant subject to the ground truth classes. Fig. 2.10(b) shows the improvement of retrieval after three rounds of RBF interactive learning. This is superior to the results obtained by MARS-2 (cf. Fig. 2.11(a)) and OPT-RF (cf. Fig. 2.11(b)). The outstanding performance of the RBF method can also be seen from Figs. 2.12-2.13, showing the retrieval results in answering the "Tiger" query. As evidenced by the results of Fig. 2.10(b) and 2.12(b), it is observed that nonlinear analysis obtained by RBF can effectively capture high-level concepts in few retrieval sessions.

The above results were obtained from 'hard' queries, which require a high degree of nonlinear discrimination analysis. There are some queries that are relatively easier to retrieve, which are shown in Fig. 2.14. Those queries have prominent features, such as a shape in the "ROSE" query, and a combination of texture and color in the "POLO" query. In each case, it is observed that MARS-2 and OPT-RF show better performance than in the previous results. In such cases, however, the retrieval results obtained by RBF approached 100% precision.

Discussion

In the past, a number of attempts have been made to describe visual contents with 'index features' for operating content-based image retrieval. The evidence

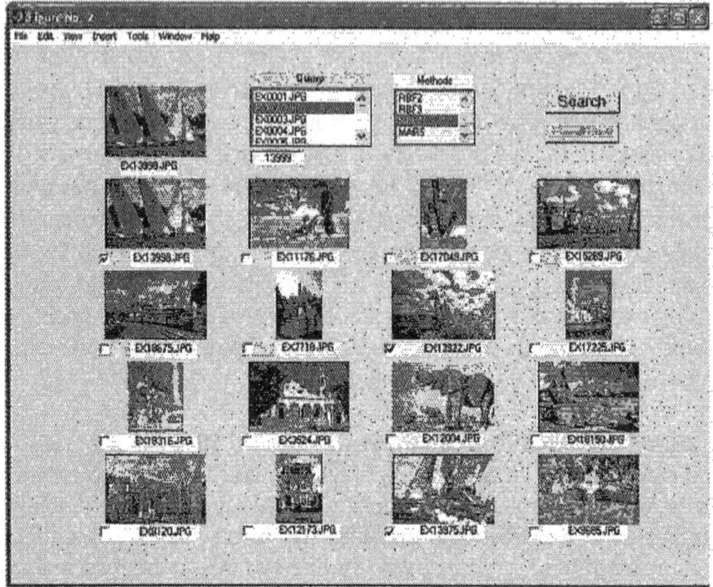

(a) Simple CBIR, precision = 0.19

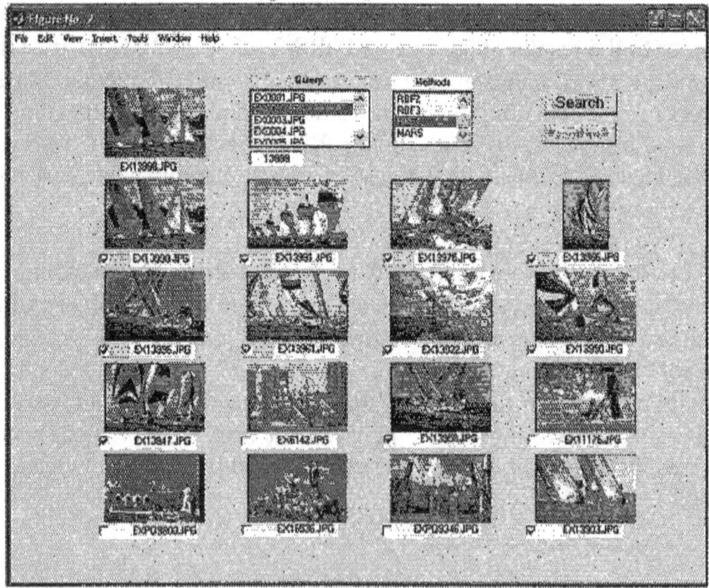

(b) RBF (t=3), precision = 0.69

Figure 2.10. Top sixteen retrieved images obtained by the "Yacht" query, using the Corel Database, (a) before RF learning, (b) after RF learning with RBF method

(a) MARS-2(t=3), precision = 0.31

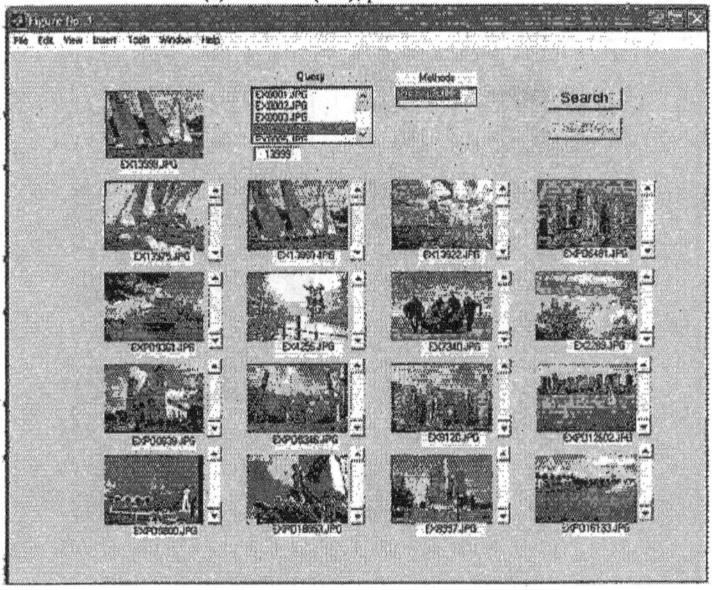

(b) OPT-RF(t=3), precision = 0.19

Figure 2.11. (*a comparison to Fig. 2.10*) Top sixteen retrieved images obtained by the "Yacht" query , using the Corel Database, after RF learning with (a) MARS-2, and (b) OPT-RF.

(a) Simple CBIR (t=0), precision = 0.19

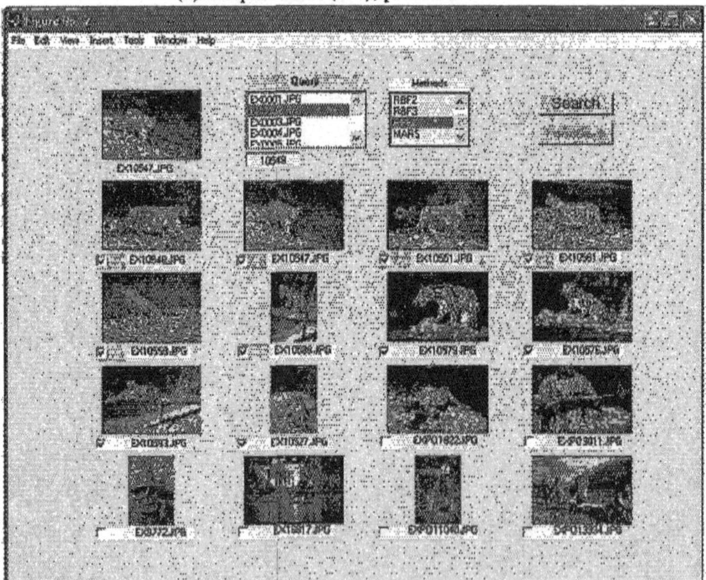

(b) RBF (t=3), precision = 0.63

Figure 2.12. Top sixteen retrieved images obtained by the "Tiger" query, using the Corel Database, (a) before RF learning, (b) after RF learning with RBF method.

(a) MARS-2(t=3), precision = 0.25

(b) OPT-RF(t=3), precision = 0.25

Figure 2.13. [*a comparison to Fig. 2.12*] Top sixteen retrieved images obtained by the "Tiger" query, using the Corel Database, after RF learning with (a) MARS-2, and (b) OPT-RF.

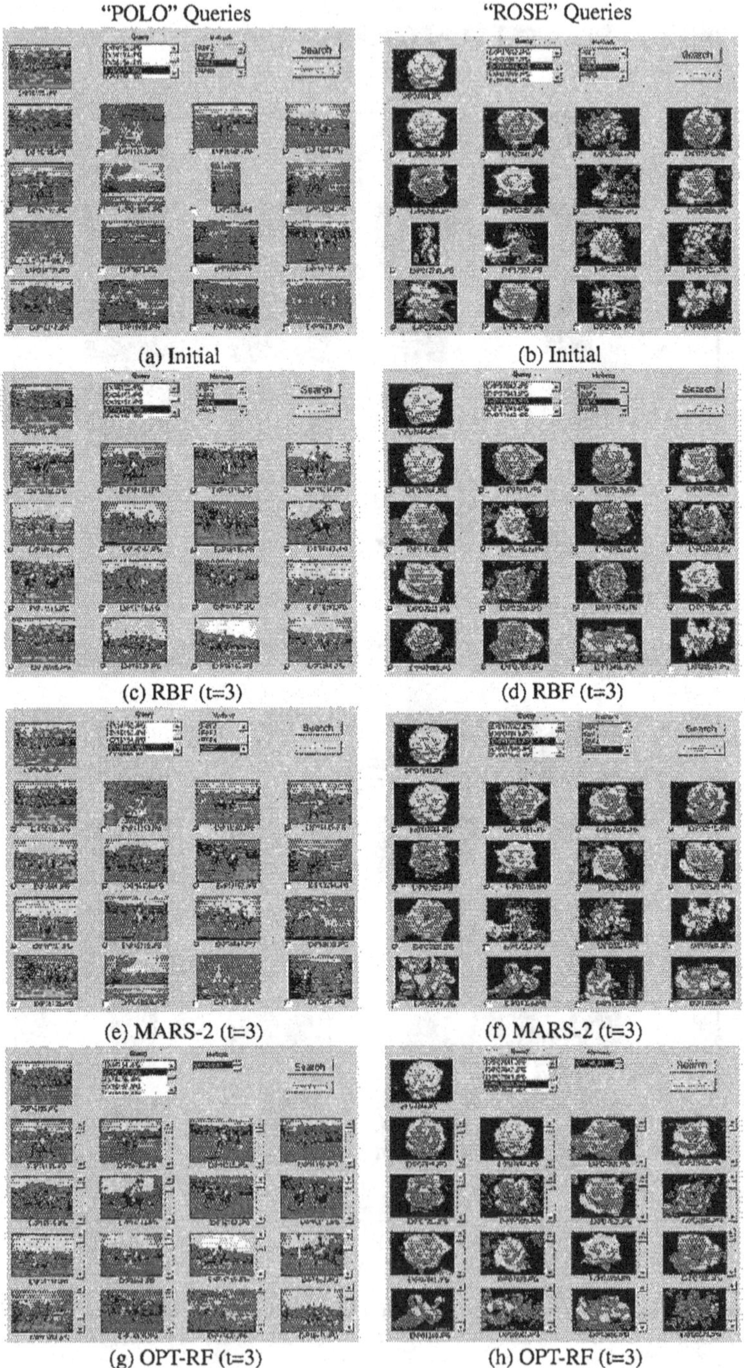

Figure 2.14. Retrieval results of "POLO" and "ROSE" queries, obtained by (a)-(b) Simple-CBIR, (c)-(d) RBF, (e)-(f) MARS-2, and (g)-(h) OPT-RF.

Color Descriptors	
Color Histogram (*d*=48) *Bins=48*	The descriptor is a 48-bin color histogram in HSV color space, where H and S are uniformly quantized into 16 and 3 regions respectively. The V component is discarded because of its sensitivity to the lighting condition [119]
Color Moments (*d*=9)	From a given RGB color image, the mean, standard deviation, and skew are extracted from the three color channels and therefore have a color feature vector of length $3 \times 3 = 9$.
Texture Descriptors	
GW transform (*d*=48)	The image is resized into 128×128 pixels in size, and converted to the gray scale level. Gabor wavelet (GW) filters spanning four scales and six orientations are then applied to the gray scale image. The mean and standard deviation of the GW coefficients are applied last to form the 48-demension feature vector.
Shape Descriptors	
Fourier Descriptors (*d*=9)	The Sobel edge detection algorithm is applied to each color channel of the RGB color image. The resulting contour edge is characterized as polar coordinate. Fast Fourier transform (FFT) is then applied to the contour edge and the coefficients in the low frequency range are truncated to form a 9-dimensional feature vector.

Table 2.2. Content descriptions used for image characterization in the Corel database.

Method	t=0	t=1	t=2	t=3	CPU time (sec./iter.)
RBF	44.82	79.82	88.75	91.79	2.34
MARS-2	44.82	60.18	61.61	61.96	1.26
OPT-RF	44.82	72.14	79.64	80.54	1.27
Simple CBIR	44.82	-	-	-	0.90

Table 2.3. Column 2-5: Average precision rate (%) obtained by retrieving 35 queries selected from different categories, using the Corel database; Column 6: Average CPU time (seconds per iteration) obtained by retrieving a single query, not including the time to display the retrieved images, using a 1.8 GHz Pentium IV processor and a MATLAB implementation.

shows that semantics and user request are more essential than 'index features' for optimum retrieval. This has directed a number of researchers to suggest that such a retrieval problem must be interpreted as human-centered, rather than computer centered [3, 4]. It has been shown in this chapter that these user information needs, in a visual-seeking environment, are well-addressed by user-interface methodologies. User interface allows the retrieval system to overcome the problem of fuzzy understanding in the user's goals, and thus aid the expression of information needs. Two main points have been demonstrated by the current method: 1) learning-based systems can adjust their strategy in accordance with user input; and 2) user information needs are satisfied by a series of selections of information.

The most difficult task in the interactive process is to analyze the role of the users in perceiving image similarity. The RBF-based interactive method has emphasized the importance of "mapping" human perception onto the image-matching process. This model incorporates and emphasizes many new features not found in earlier interactive retrieval systems. Many of these features are imparted by *nonlinear* discriminant analysis with a high degree of adaptivity through learning from negative and positive samples. This results in a high performance learning machine that learns effectively and quickly from a small set of feedback data. It has been suggested that through a learning-based approach it is possible to relate the behavior of human perception to low-level feature processing in visual retrieval systems. The learning-based approach takes into account the complexities of individual human perception and, in fact, uses individual user choices to decide relevance. This learning machine combines state-of-the-art retrieval performance with a very rich set of features, which may help to usher in a new generation of multimedia applications.

7. A Computer Aided Referral (CAR) System for Mine Target Detection in Side-Scan Sonar Images

Introduction

Computer aided detection (CAD) method has found many potential applications in our society, especially in defense and in medicine, such as the detection of underwater mine-like objects in side-scan sonar images and the detection of breast cancer in digital mammograms.

In the detection of underwater mine-like objects, the CAD method points to particular areas in a sonar image and label the areas as "potentially dangerous". Because of technological limitations imposed by state-of-the-art in image processing and pattern recognition, a significant percentage of the positive cases are not detected by the CAD systems in time for early actions. Similar problems have been observed in the detection of breast cancer: about 10% of the malignant cases are not detected in time for early treatment.

In light of the problems associated with the conventional CAD methods, it is proposed to investigate an alternative way, applying content-based image retrieval (CBIR) techniques to assist operators/doctors. The CBIR techniques take a computer aided referral (CAR) approach based on the content information embedded in the images. This section will present the work on the development of a CAR system for the detection of underwater mine-like objects in side-scan sonar images.

Underwater mines may cause serious problems for surface and submarine vessels. Side-scan sonar has been recognized as an effective way of detecting mine objects. However, human operators must monitor the images collected by the sonar, and targets might be missed due to inconsistencies in performance. Since the underwater environment is normally very complicated with noise and background clutter of various kinds, it is very important to identify the mine objects in the sonar images so that the operator can better correlate the detection results with the potential targets visually observed. In addition, an effective detection method is a vital step towards the recognition of the mines.

In the CAR approach, a digital library is first established which contains sonar images with signatures of various mine objects, and objects with signatures similar to mines. In operation, when a suspicious object is observed in a captured sonar image, this image will be compared with the images stored in the sonar image library based on their visual characteristics (features). Then the N images with the most similar characteristics to the captured image are brought to the attention of the operator. The operator will compare the captured image with those retrieved from the library. If the captured image indeed has similar visual features to one or more images with mines in the library, this image will be more carefully studied by the operator in order to make an inferred decision of the case on hand. Because the CAR approach takes a much broader range of possible characteristics of the mine signatures into consideration, instead of simply attempting to match to a "standard" template as a CAD method does, it can potentially minimize the high miss rate experienced by the CAD method.

Brief Description and Summary of the Preliminary Study

In a preliminary study, a library of sonar images was received from Defense Research and Development Canada (DRDC) Atlantic. The library contains 383 images with a variety of targets. Some of the images also have multiple targets. Fig. 2.15 show images of some of the targets. There are 10 different types of mines in these images. A CAR system is developed to aid the mine detection. This is a semi-automatic approach where we short list the number of probable classes of the target to N ($N \leq 3$) using the content-based image retrieval techniques. The operators can then further investigate the short listed images for faster and more accurate detection of mines. In general, a CAR mine detection system consists of three stages, enhancing/denoising the images, locating the targets (both mine and non-mine objects) in the images, and detecting the mines.

In operation, the regions of interest was extracted from the sonar images in the database. These cropped images are then used for extraction of the features for the classification. The system consists of the following modules:

- Preprocessing Module
- Feature Extraction Module
- Search Engine based on linear similarity measurement
- Graphical User Interface (GUI)

The image data is read from the files in the image library first. The preprocessing module is then used to convert the data into the required format and extract the region of interest from the images. Next the data is fed to the feature extraction module. This module extracts 29 statistical features representing each object in the sonar image. These features are then used by the search engine to find the top N matches. MatLab and SPSS were used for the implementation of the analysis and classification modules of the system. The system is designed in such a way that a user can present a sonar image for analysis through the GUI. The system extracts the features from the image and gives the N most probable classes of the sonar images in the image library to which the input image might belong. The user can then further investigate and find if the signature in the captured image is indeed that of a mine. It has been observed that, when setting $N = 2$, the detection rate is 77.29% with a false positive rate of 17.81% . If $N = 3$ was set, the detection rate is substantially increased to 95.12% , but the false positive rate is also high, over 40% .

Discussions

Our preliminary results demonstrated that the CAR approach is potentially an effective method in detecting underwater mine-like objects in side-scan sonar images. The more attractive result is observed when $N = 3$ with 95.12% detection rate. Although the false positive rate is also high, the key is that only very few mine signatures are missed. However, the high false positive rate has to be addressed before the system could be useful. In fact, the high false positive rate is easy to explain: the set of data samples is too small, especially the samples in the mine classes. Since the CAR system is based on image retrieval, performance of which is critically dependent on the comprehensiveness of the image library and the quality of the features. With less than 90 samples, the mine classes are highly under-represented. Therefore, the foremost important issue to reduce the false positive rate while keeping the detection rate high is to have a large sonar image library which contains mine samples with all the possible characteristics, e.g. different scales, different angles of rotations, etc. In addition, the following steps should also improve the performance the CAR system.

- Currently only low level texture features are used in retrieval. Exploring more effective features such as the newly developed wavelet modeling by mixture of Laplacians, and shape-based features will certainly enhance the performance of the system.
- A linear function is used for similarity matching in the CAR system. We can apply more advanced similarity measurements such as the RBF networks or support vector machines to improve the pattern recognition module of the system.
- Relevance feedback is an effective means to improve retrieval performance. Incorporating automatic relevance feedback techniques will improve the retrieval accuracy without adding burden to the operators.
- Some sonar images are subject to different kinds of noise. Denoising can help to improve the quality of the images and lead to better detection performance.

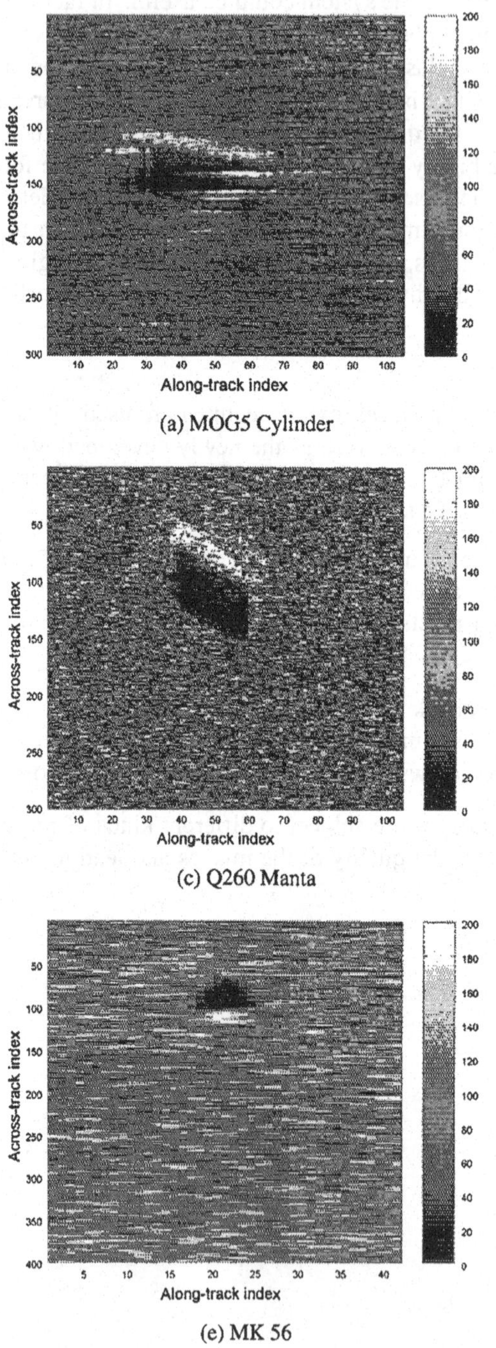

Figure 2.15. Sindescan-sonar images of some of the targets

Chapter 3

HIGH PERFORMANCE RETRIEVAL: A NEW MACHINE LEARNING APPROACH

1. Introduction

In image retrieval, particularly in general image collections, the relevancy of images to a specific query is most appropriately characterized by a multiple-class modeling approach. For example, when a user has a query for a "PLANE," she or he may wish to have any image containing planes, as shown in Fig. 3.1. The *semantic* of "PLANE" is clearly described by a variety of models, which are correlated, but each of which has its own local characteristics. The difficulty in characterizing image relevancy, then, is identifying the local context associated with each of the "sub-classes" within the class "PLANE." Human beings utilize multiple types of modeling information to acquire and develop their understanding about image similarity. To obtain more accurate, robust, and natural characterizations, a computer must generate a fuller definition of what humans regard as significant features. Through Human-Computer Interaction (HCI), computers do acquire knowledge of novel features which are significant but have not been explicitly specified in the training data. This implicit information constitutes subclasses within the query, permitting better generalization. In this chapter, a mixture of Gaussian models is used, via the RBF network, to represent multiple types of model information for the recognition and presentation of images by human beings and machines.

Previously, Chapter 2 introduced a nonlinear input-output mapping function based on a single RBF model. As with most other works [6, 8, 11] discussed, this has been concerned with 'global' modeling, in which a query image is described by one model, which is then associated with only a particular location in the input space. Furthermore, the similarity function is based on a single matrix.

The combination gives rise to a single model function, $f(\mathbf{x}, \mathbf{z})$, which cannot fully exploit the local data information. This chapter introduces a mixture of Gaussian models for interactive retrieval that enables the learning system to take advantage of the information from multiple sub-classes. The proposed learning system utilizes a highly local characterization of image relevancy in the form of a superposition of different local models, as $\sum f(\mathbf{x}, \mathbf{z})_i$, to obtain the input-output mapping function. As a result, local data distributions can be sufficiently exploited to achieve rapid performance improvements in retrieval.

The development of new learning strategy for RBF network is introduced for interactive retrieval, which is based on the local model network architecture proposed by Poggio and Girosi [81]. The learning system produces Gaussian classifications that represent the statistics of the input data, and the system uses a learning scheme of moving centers and adapting weighted norms to update these pre-developed classes. This is in contrast to previous uses of the RBF network for a single model, as in Chapter 2. The primary motivation here is to specify sufficient units for the system to recognize a relevant image defined by different viewpoints. Within a neural computing approach, retrieval approaches human capability.

2. Local Model Networks (LMN)

The basic assumption underlying the use of learning systems is that the behavior of the system can be described in terms of the training set $\mathcal{T} = \{(\mathbf{x}_1, y_1), ..., (\mathbf{x}_N, y_N)\}$. It is therefore assumed that the system to be described by a model whose observable output y_i, at the time of step i, in response to an input vector $\mathbf{x}_i$, is defined by:

$$y_i = f(\mathbf{x}_i) + \varepsilon_i, \quad i = 1, 2, ..., N \tag{3.1}$$

where ε_i is a sample drawn from a white noise process of zero mean and variance σ^2.

The modeling problem is to estimate the underlying function of the model, $f(\mathbf{x}_i)$, from observation data, having already used the existing *a priori* information to structure and parameterize the model $\hat{f}$. Let $\hat{f}(\mathbf{x}, \mathbf{z})$ be the estimate of $f(\mathbf{x})$ for some values of the P-dimensional parameter vector $\mathbf{z}$. In general, the model structure cannot exactly describe the system, so a bias $b(\mathbf{x})$ is naturally associated with $\hat{f}$:

$$y = \hat{f}(\mathbf{x}, \mathbf{z}^*) + b(\mathbf{x}) + \varepsilon \tag{3.2}$$

Figure 3.1. Example images representing semantic of "PLANE".

The optimal $\mathbf{z}^*$ in the learning problem is defined such that the average bias is minimized over the entire data set:

$$\mathbf{z}^* = \arg\min_{\mathbf{z}} \sum_{i=1}^{N} \left(f(\mathbf{x}_i) - \hat{f}(\mathbf{x}_i, \mathbf{z}) \right)^2 \tag{3.3}$$

Since f is not known, these optimal parameters cannot be obtained. We, therefore, look for an estimate based on the training data $\{(\mathbf{x}_i, y_i)\}_{i=1}^{N}$.

The model function $\hat{f}$ can be estimated in a number of ways. As discussed in Chapter 2, for a single model, $\hat{f}$ may be constructed by a supervised learning for the feature weighing scheme and/or a reformulation of a query model. In the classification context, this model is viewed as a single cluster in the feature space, as shown in Fig. 3.2(a). Such a model yields a *global* characterization of image similarity. For multiple models, however, it may be an advantage to partition the input space into multiple subspaces [cf. Fig. 3.2(b)]—an inherent strategy of local modeling techniques. A *Local Model Network* (LMN) [cf. Fig. 3.3] is therefore adopted to achieve this purpose [81, 87]. This type of network approximates the model function $\hat{f}$ according to:

$$\hat{f}(\mathbf{x}, \mathbf{z}) = \sum_{i=1}^{n_M} \lambda_i \hat{f}_i(\mathbf{x}, \mathbf{z}_i) \tag{3.4}$$

$$\equiv \sum_{i=1}^{n_M} \lambda_i \phi_i(||\mathbf{x} - \mathbf{z}_i||) \tag{3.5}$$

where $\mathbf{x} = [x_1, ..., x_P]^T \in \mathcal{R}^P$ and $\mathbf{z} = [z_1, ..., z_P]^T \in \mathcal{R}^P$ are the input vector and the RBF center, respectively; $\lambda_i, i = 1, ..., n_M$ are the weight; and $\phi(\cdot)$ is a nonlinear function from $\mathcal{R}^+$ to $\mathcal{R}$, referred to as nonlinearity of hidden nodes.

The advantage of this network's use in the current application is that it finds the input-to-output map using local approximators; consequently, the underlying basis function responds only to a small region of the input space where the function is centered, e.g., a Gaussian response, $\phi(d) = \exp(-d^2/2)$, where:

$$d(\mathbf{x}, \mathbf{z}_i, \sigma_i) = \sqrt{(\mathbf{x} - \mathbf{z}_i)^T \sigma_i^{-2} (\mathbf{x} - \mathbf{z}_i)}. \tag{3.6}$$

This allows local evaluation for image similarity matching.

Typically, the parameters to learn for the LMN are the set of linear weight λ_i, the center $\mathbf{z}_i$, and the width σ_i for each local approximator, $\phi_i, i \in \{1, ..., n_M\}$. The linear weights are usually estimated by using the least-square (LS) method [82]. When using the Gaussian function as the nonlinearity of hidden nodes, it has been observed that the same width of σ_i is sufficient for the RBF network to obtain universal approximation [84]. However, more recent theoretical investigations and practical results indicate that the choice of center $\mathbf{z}_i$ is most significant in the performance of the RBF network [83]. As we shall see, this suggestion plays a central role in overcoming the variation of the performance of the network in the interactive retrieval learning application.

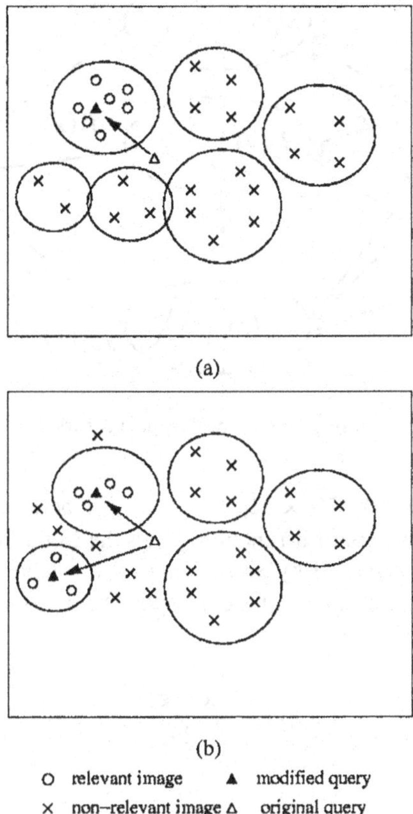

Figure 3.2. Illustration of data clustering and query characterization using (a) a global model, and (b) using a local model.

In the following sections a brief review of the standard solution to interactive learning in image retrieval is given, using the LS method [82] and the orthogonal LS learning algorithm [83]. Then, Section 5 will describe a new alternative learning approach that is more suitable for learning with a small training set.

3. Learning via Local Model Network

Various learning strategies have been proposed to structure and parameterize the RBF network [13, 81–83]. This section will consider two of these beside the new learning strategy for interactive image retrieval.

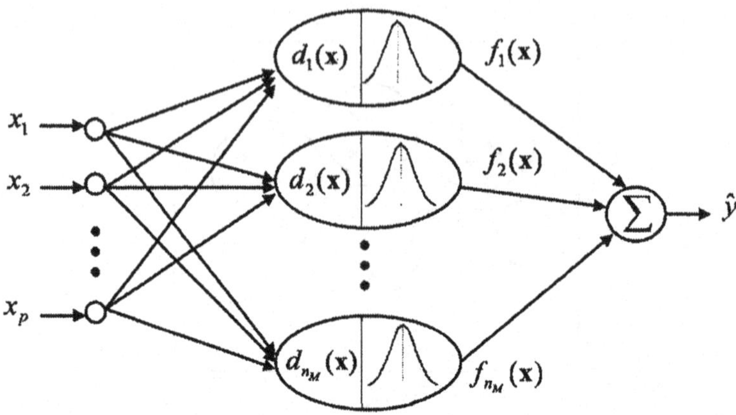

Figure 3.3. Local model network.

Let $\mathcal{T} = \{(\mathbf{x}_i, y_i) \in \mathcal{R}^P \times \mathcal{R} | \, i = 1, ..., N\}$ be a set of data we intend to approximate by means of a function $\hat{f}$. In the initial approaches [81], the RBF network is constructed by associating all available training samples to the hidden units, using one-to-one correspondence. A radial-basis function centered at $\mathbf{z}_i$ is defined as:

$$\phi(||\mathbf{x} - \mathbf{z}_i||) = \exp\left(-\frac{||\mathbf{x} - \mathbf{z}_i||^2}{2\sigma_i^2}\right), i = 1, 2, ..., n_M \quad (3.7)$$

where

$$\{\mathbf{z}_i\}_{i=1}^{n_M} = \{\mathbf{x}_i\}_{i=1}^{N}; \quad n_M = N \quad (3.8)$$

This solution, however, proved to be expensive, in terms of computational complexity, because N is large.

A method proposed by Brommhead and Lowe [82], is to arbitrarily choose some data points as centers. This gives an approximation to the original RBF network, while providing a more suitable basis for practical applications. In this case, the number of basis functions is less than the number of training data points (i.e., $n_M \leq N$), and the approximated solution $\hat{f}^*(\mathbf{x})$ is expanded on a finite basis:

$$\hat{f}^*(\mathbf{x}) = \sum_{i=1}^{n_M} \lambda_i \phi_i(\mathbf{x}, \mathbf{z}_i) \quad (3.9)$$

where

$$\{\mathbf{z}_i\}_{i=1}^{n_M} \subset \{\mathbf{x}_i\}_{i=1}^{N} \quad (3.10)$$

The linear weights $\lambda = [\lambda_1, ..., \lambda_i, ..., \lambda_{n_M}]^T$ are determined by minimizing the cost functional $\xi(\hat{f}^*)$, thus:

$$\xi(\hat{f}^*) = \sum_{i=1}^{N} \left(y_i - \sum_{j=1}^{n_M} \lambda_j \phi_j \left(\|\mathbf{x}_i - \mathbf{z}_j\| \right) \right)^2 + \gamma \left\| \mathbf{D} \hat{f}^* \right\|^2 \tag{3.11}$$

where γ is the regularization parameter, and $\mathbf{D}$ is a differential operator. Based on the *pseudoinverse method* [82], the minimization of Eq.(3.11) with respect to the weight vector $\lambda = [\lambda_1, ..., \lambda_i, ..., \lambda_{n_M}]^T$, yields:

$$\lambda = \mathbf{G}^+ \mathbf{y} \tag{3.12}$$

$$= \left(\mathbf{G}^T \mathbf{G} \right)^{-1} \mathbf{G}^T \mathbf{y} \tag{3.13}$$

where

$$\mathbf{y} = [y_1, y_2, ..., y_N]^T \tag{3.14}$$

The matrix $\mathbf{G} \in \mathcal{M}_{N \times n_M}$ is defined as:

$$\mathbf{G} = \{\phi_{ij}\} \tag{3.15}$$

$$\phi_{ij} = \exp\left(-\frac{\|\mathbf{x}_i - \mathbf{z}_j\|^2}{2\sigma_j^2} \right), \quad i = 1, 2, ..., N; \; j = 1, 2, ..., n_M \tag{3.16}$$

where $\mathbf{x}_i \in \mathcal{T}$ is the ith input vector of the training samples.

This method is referred to as the randomly selected center-RBF network (RC-RBF). The main problem with this method is that it cannot guarantee desired performance, because it may not satisfy the requirement that the centers should suitably sample the input domain. In some cases, a "large" RBF network may be required to achieve the preferred performance. Taking this problem into account, the orthogonal least squares (OLS) learning algorithm is proposed by [83], to select a suitable set of centers so that adequate and parsimonious RBF networks can be obtained. The OLS algorithm chooses centers one by one from the training data; that is, at each iteration the vector that results in the largest reduction in network errors is used to create the center. When the sum-squared error of the network computed is higher than a specified level, the next center is added to the network. The iteration process stops when the error falls beneath an error goal, or when the maximum number of centers is reached. This provides a simple and efficient means for fitting RBF networks [83].

4. Machine Learning-Based Interactive Retrieval

The learning strategies described above are generally designed to construct the RBF network to achieve *universal approximation*, using a large number of training data. For interactive image retrieval, however, the characteristic of learning is quite different. First, the training data size for image retrieval is very small compared to the general approximation strategy. Second, the training samples available for image retrieval are highly correlated, i.e., each sample is selected from a specific area of the input space and is near to the next, in the Euclidean sense. For a general function approximation, the training data is expected to cover a larger area of the input space, and so is usually selected in a random manner. Therefore, the following definition is made:

DEFINITION 1 *Let function $f(\mathbf{x})$ described by Eq.(3.9) be an approximation function for image retrieval, which is used to identify relevant images from a database. For any given input image feature vector $\mathbf{x}$, the retrieval system will return a (normalized) scalar output $y = f(\mathbf{x}) \cong 1$ if $\mathbf{x}$ is in the relevant class, otherwise $y \cong 0$.*

Let us focus on a given example for which the function $f(\mathbf{x})$ is estimated. Assume that a set $\mathcal{T}_q$ contains a relevant sample $\mathbf{z}_r$ and a non-relevant sample $\mathbf{z}_{ir}$. Without loss of generality, it is further assumed that the sample set in $\mathcal{T}_q$ is used to estimate two RBF functions: $\phi_r(\mathbf{x}, \mathbf{z}_r)$ and $\phi_{ir}(\mathbf{x}, \mathbf{z}_{ir})$. Thus, according to the learning strategies Eq.(3.12), there are two linear weights associated with the RBF functions. Since $\phi_r(\mathbf{x}, \mathbf{z}_r)$ represents the relevant class, its corresponding linear weight λ_r, when computed, is positive and greater than the weight λ_{ir}, which corresponds to $\phi_{ir}(\mathbf{x}, \mathbf{z}_{ir})$; that is:

$$\begin{aligned} \lambda &= [\lambda_r, \lambda_{ir}]^T \\ \lambda_r &> 0 > \lambda_{ir} \end{aligned} \tag{3.17}$$

We may then form this approximating function $f(\mathbf{x})$:

$$f(\mathbf{x}) = \lambda_r \phi_r(\|\mathbf{x} - \mathbf{z}_r\|) + \lambda_{ir} \phi_{ir}(\|\mathbf{x} - \mathbf{z}_{ir}\|) \tag{3.18}$$

which is the superposition of two Gaussian-shaped RBF functions centered at $\mathbf{z}_r$ and $\mathbf{z}_{ir}$.

The estimated function $f(\mathbf{x})$ is then used for the classification. The following cases examine the performance of $f(\mathbf{x})$ in two different environments: Case 1 for a learning problem in interactive retrieval; and Case 2 for a general function approximation problem.

Let $\mathbf{x}'$ be an input vector that is already known as a "relevant" sample.

Case 1: If $\mathbf{z}_r$, $\mathbf{z}_{ir}$, and $\mathbf{x}'$ are located very close to each other in a small region $R \subset \mathcal{R}^P$, the negative weight λ_{ir} assigned to reduce the effect of $\phi_{ir}(\mathbf{x}, \mathbf{z}_{ir})$ around $\mathbf{z}_{ir}$, also reduces the positive weights constructed within the small region R, i.e., it reduces the contribution of $\phi_r(\mathbf{x}, \mathbf{z}_r)$ towards constructing $f(\mathbf{x}')$. As a result, the output $f(\mathbf{x}')$ obtained from the superposition of $\phi_r(\mathbf{z}_r, \mathbf{x}')$ and $\phi_{ir}(\mathbf{z}_{ir}, \mathbf{x}')$, has a low magnitude; and so fails to retrieve the relevant sample $\mathbf{x}'$. This may also result in lowering the possibility of retrieving other relevant samples (which have not yet been retrieved) from this small region R.

Case 2: If $\mathbf{z}_r$ and $\mathbf{z}_{ir}$ are not in the neighborhood area, the negative weight λ_{ir} has less effect on the magnitude of $\phi_r(\mathbf{z}_r, \mathbf{x}')$. Thus $\phi_r(\mathbf{z}_r, \mathbf{x}')$ contributes more to the classification of $\mathbf{x}'$ into the correct class, providing that $\mathbf{x}'$ is close to $\mathbf{z}_r$.

In other words, when the training samples are highly correlated, the choice of centers is the most important factor for the effectiveness of constructing function approximation. Particularly, the local approximators estimated by the non-relevant samples result in reduced effectiveness of the overall performance. In the learning context of the RBF network [83], this problem can be understood as being "ill-conditioned", owing to the near-linear dependency caused by "some centers being too close together".

5. Adaptive Local Model Network

In order to circumvent the environmental restrictions in Case 1, an adaptive learning strategy for the RBF network is introduced and referred to as adaptive RBF network (ARBFN). This is a special network for learning in image retrieval where there is a small set of samples with a high level of correlation between the samples. This new strategy is based on the following points:

- Formulate and solve the local approximator $\phi(\cdot)$ from available *positive* samples
- In order to take advantage of negative samples to improve decision boundary, a method of shifting centers is obtained, instead of employing linear weights.
- In order to obtain a dynamic weighting scheme, the Euclidean norm in $\phi(\|\mathbf{x} - \mathbf{z}_i\|)$ is replaced with the weighted Euclidean, which provides a new form of local approximator, $\phi(\|\mathbf{x} - \mathbf{z}_i\|_\Lambda)$, where adjustable metric parameters are estimated according to the selection of relevant images.

The learning strategy for the ARBFN consists of two parts. First, the local approximators $\phi(\cdot)$ and their associated centers and widths are constructed using positive samples. Second, in order to improve the decision boundary, negative samples are used for shifting the centers, based on anti-reinforced learning [85]. These are described in the next two sections.

Estimation of Local Approximators

Let $\mathcal{X}^+ = \{(\mathbf{x}'_i, y_i) | \, y_i = 1, \, i = 1, 2, ..., n_M\}$ be a set of the positive samples obtained from the training data set $\mathcal{T}$, $\mathcal{X}^+ \subset \mathcal{T}$. In order to construct the local approximators $\phi(\cdot)$, the ARBFN learning algorithm uses these positive samples to estimate the RBF centers and widths. Each positive sample is assigned to the local approximator $\phi(\cdot)$, so that the shape of each relevant cluster can be described by:

$$\phi(||\mathbf{x} - \mathbf{z}_i||) = \exp\left(-\frac{||\mathbf{x} - \mathbf{z}_i||^2}{2\sigma_i^2}\right) \tag{3.19}$$

where the centers $\mathbf{z}_i$, $i = 1, 2, ..., n_M$ are characterized by the positive samples

$$\{\mathbf{z}_i\} = \{\mathbf{x}'_i\}, \; i = 1, 2, ..., n_M \tag{3.20}$$

The RBF width σ_i used in Eq.(3.19) is estimated by

$$\sigma_i = (\delta) \min_j \left(\left\| \mathbf{x}'_i - \mathbf{x}'_j \right\| \right), \; \forall \mathbf{x}'_j \in \mathcal{X}^+, \; j \neq i \tag{3.21}$$

where $\delta = 0.5$ is an overlapping factor.

Here, only the positive samples are assigned as the centers of the corresponding RBF functions. To take into account the effects of the negative samples, a method of shifting centers is introduced (described in Section 5.0). Hence, the estimated model function $f(\mathbf{x})$ is given by:

$$f(\mathbf{x}) = \sum_{i=1}^{n_M} \lambda_i \phi_i(||\mathbf{x} - \mathbf{z}_i||), \tag{3.22}$$

$$\lambda_i = 1, \; \forall i = 1, .., n_M \tag{3.23}$$

The linear weight, $\lambda_i = 1$, indicates that all the centers (or the positive samples) are taken into consideration. However, the degree of importance of $\phi_i(\cdot)$ is indicated by the natural responses of the Gaussian-shaped RBF functions and the superposition of the functions. For instance, if centers $\mathbf{z}^{(a)}$ and $\mathbf{z}^{(b)}$ are highly correlated (i.e., $\mathbf{z}^{(a)} \approx \mathbf{z}^{(b)}$), the magnitude of $f(\mathbf{x})$ will be biased for any input vector $\mathbf{x}$ located near $\mathbf{z}^{(a)}$ or $\mathbf{z}^{(b)}$, i.e., $f(\mathbf{x}) \approx 2\phi(\mathbf{x}, \mathbf{z}^{(a)}) \approx 2\phi(\mathbf{x}, \mathbf{z}^{(b)})$. This naturally increases the positive weight for the classification.

As described by Eq.(3.22), the ARBFN performs the similarity evaluation by linearly combining the output $\phi_i(\cdot)$. If the current input vector $\mathbf{x}$ (associated with an image in the database) is close to one of the RBF centers $\mathbf{z}_i$ in the Euclidean sense, the corresponding RBF unit $\phi(\mathbf{x}, \mathbf{z}_i)$ will also increase, indicating this greater similarity of the corresponding image. On the other hand, input vectors that are far away from each of $\mathbf{z}_i$, $i = 1, 2, ..., n_M$ do not appreciably contribute to the summation due to the exponentially decaying weighting function.

A Weighted Norm

The basic RBF version of the ARBFN discussed above (Eq.(3.19)) is based on the assumption that the feature space is uniformly weighted in all directions. However, as discussed in Chapter 2, image feature variables tend to exhibit different degrees of importance which heavily depend on the nature of the query and the relevant images defined [6]. This leads to the adoption of an elliptic basis function (EBF):

$$\psi(\mathbf{x}, \mathbf{z}_i) = \|(\mathbf{x} - \mathbf{z}_i)\|_{\Lambda}^2 = (\mathbf{x} - \mathbf{z}_i)^T \mathbf{\Lambda} (\mathbf{x} - \mathbf{z}_i) \tag{3.24}$$

where

$$\mathbf{x} = [x_1, ..., x_P]^T \in \mathcal{R}^P \tag{3.25}$$

$$\mathbf{z}_i = [z_{i1}, ..., z_{iP}]^T \in \mathcal{R}^P \tag{3.26}$$

$$\mathbf{\Lambda} = diag[\alpha_1, ..., \alpha_p, ..., \alpha_P] \tag{3.27}$$

So, the parameters α_p, $p = 1, ..., P$ represent the relevance weights which are derived from the variance of the positive samples in $\mathcal{X}^+ = \{(\mathbf{x}_i', y_i) | \; y_i = 1, i = 1, 2, ..., n_M\}$ as follows:

$$\alpha_p = \begin{cases} 1 & \xi_p = 0 \\ 1/\xi_p & \text{otherwise} \end{cases} \tag{3.28}$$

where

$$\xi_p = \left(\frac{1}{n_M - 1} \sum_{i=1}^{n_M} (x_{ip}' - \bar{x}_p')^2 \right)^{\frac{1}{2}} \tag{3.29}$$

$$\bar{x}_p' = \frac{1}{n_M} \sum_{i=1}^{n_M} x_{ip}' \tag{3.30}$$

(3.31)

The matrix $\mathbf{\Lambda}$ is a symmetrical $\mathcal{M}_{P\times P}$, whose diagonal elements α_p assign a specific weight to each input coordinate, determining the degree of the relevance of the features. The weight α_p is inversely proportional to ξ_p, the standard deviation of the sequence $\{x'_{ip}|\ i = 1, 2, ..., n_M\}$. This weighting criterion is based on the assumption that if a particular feature is relevant, then all positive samples should have a very similar value to this feature, i.e., the sample variance in the positive set is small [8]. This adjustable metric is very important, since it permits a more specific dissimilarity measure between inputs and centers than Euclidean metric.

As a result, a new version of the Gaussian-shaped RBF Eq.(3.19), that takes into account the feature relevancy, can be defined as:

$$\phi(||\mathbf{x} - \mathbf{z}_i||_{\Lambda}) = \exp\left(-\frac{\psi(\mathbf{x}, \mathbf{z}_i)}{2\sigma_i^2}\right) \tag{3.32}$$

where $\psi(\mathbf{x}, \mathbf{z}_i)$ is as given in Eq.(3.24).

Shifting Centers

The possibility of moving the expansion centers is useful for improving the *representativeness* of the centers. Recall that, in a given training set, both positive and negative samples are presented, which are ranked results from the previous search operation. Let $\mathcal{X}^- = \{(\mathbf{x}''_i, y_i)|\ y_i = 0, i = 1, 2, ..., n_{ir}\}$ be the set of negative samples. For all negative samples in this set, the similarity scores from the previous search indicate that their clusters are close to the positive samples retrieved. Here, the use of negative samples becomes essential, as the RBF centers should be moved slightly away from these clusters. Shifting the centers reduces the similarity scores for those negative samples, and thus more favorable similarity scores can be obtained for any positive samples that are in the same neighborhood area, in the next round of retrieval. Therefore, each node vector is modified in the positive sample set, $\mathcal{X}^+ = \{\mathbf{x}'_i|\ i = 1, 2, ..., n_M\}$, before it can be used for characterizing the RBF centers in Eq.(3.20).

Let the input vector $\mathbf{x}''$ (randomly selected from the negative data set) be the closest point to $\mathbf{x}'_{i^*}$, such that:

$$D_{i^*} < D_i; \quad \forall i \in \{1, 2, ..., n_M\}, \quad i \neq i^* \tag{3.33}$$

where D_i is the weighted distance:

$$D_i = \sum\nolimits_{p=1}^{P} \alpha_p \left(x''_p - x'_{ip}\right)^2 \tag{3.34}$$

Then, the node vector is modified by the anti-reinforced learning rule:

$$\mathbf{x}'_{i^*}(n+1) = \mathbf{x}'_{i^*}(n) - \eta(n)[\mathbf{x}''(n) - \mathbf{x}'_{i^*}(n)] \tag{3.35}$$

where $\eta(n)$ is a learning constant which decreases monotonically with the number of iterations n, $0 < \eta(n) < 1$. The algorithm is repeated by going through the collection of samples $\{\mathbf{x}''_i,\ i = 1, 2, ..., n_{ir}\}$.

6. Application to Corel's Photograph Collection

A number of experiments were conducted to evaluate the LMN for image retrieval. For this experiment, the Corel database as used, as in the experiments reported in Chapter 2 (Section 2.6). That is, all 40,000 images in the database were used, each of which was characterized by a multi-feature representation including shape, color, and texture. This section begins by implementing the LMN using the ARBFN architecture and comparing its performance with two other learning strategies. This is followed by examining the ARBFN and the global models discussed in Chapter 2.

Retrievals with local model networks

Our purpose in this section is to verify that the proposed ARBFN is able to meet the demands of interactive retrieval application; in particular, where there is a small set of training samples with a high level of correlation between the samples. A learning session with this condition may be observed in Fig.3.10, where the top sixteen retrieved images are returned to the user who provides relevance feedback. It is seen that at later iterations the learning system can improve the result sets, which means that the more times the interactive retrieval is implemented, the higher the level of correlation among the retrieved images.

The ARBFN method was compared with two learning strategies that have been successfully used in other situations to construct the RBF network. Both learning strategies have been implemented and are available from the Neural Network Toolbox using MATLAB [87]. The first learning method[1], the orthogonal least square (OLS) learning procedure described in [83], was used to identify a RBF network model. The RBF centers were chosen one at a time from a retrieved vector set. The selection started with zero centers, and new centers were iteratively picked in the subsequent selection procedure. Each time, the network's mean square error was checked and compared to the pre-defined tolerance set at 0.0001. The algorithm terminated either when the maximum number of centers was reached (sixteen), or when the network's mean square error was lower than 0.0001. In the second learning method[2], each vector in a retrieved set was associated with the RBF centers (Eq.(3.8)), using a one-to-one correspondence. The weight and bias of the second layers were calculated in such a way that the network sum-squared error was minimized to zero on the training vectors. This method is denoted as EDLS (Exact Design Network using Least Squares criterion). The final RBF network model can be written as:

$$\hat{y}(\ell) = \lambda_0 + \sum_{i=1}^{n_M} \lambda_i \phi(\|\mathbf{x}(\ell) - \mathbf{z}_i\|) \tag{3.36}$$

where $n_M = 16$ for the EDLS, and $n_M \leq 16$ for the OLS learning method. The input vector $\mathbf{x}(\ell)$ was used to evaluate the degree of similarity of the corresponding image, and the resulting scores from the entire database, $\hat{y}(\ell), \ell = 1, 2, ..., 40000$ were ranked in descending order to obtain the top sixteen retrievals. The RBF widths for both learning methods were determined experimentally, and it was found that the appropriate width was $\sigma = 0.8$.

In the proposed learning algorithm, the ARBFN described in Section 5 was used to identify an RBF network model. The positive samples were associated with the RBF centers, and the negative samples were used to modify the centers. The RBF widths and the weighted norm parameters were dynamically adjusted by the positive samples, using Eq.(3.21) and Eq.(3.32) respectively.

Thirty five images were chosen as queries, from different categories [cf. Fig. 2.9]. The query set used here is identical to the experiments reported in Chapter 2 (Section 2.6). For each query, relevance feedback (subjected to The ground truth classes) was provided on the 16 top-ranked images retrieved, and the interactive process was conducted through three iterations. Precision (Pr) was recorded after each query iteration.

Table 3.1 summarizes the average precision results, $\bar{Pr}_{(t)}$, as a function of iteration t, taken over the 35 test queries. It can be seen from the results that the ARBFN significantly improved the retrieval accuracy (up to 92% precision). The first iteration showed an improvement of about 35.9%; the second iteration an additional 9.6% , and the third increased by 2.1% . The ARBFN outperformed the OLS (76.61%) and the EDLS. This result confirms that the ARBFN learning strategy offers a better solution to the construction of an RBF network for the interactive image retrieval, than the two standard learning strategies.

Both the OLS and the EDLS strategies usually perform well under the opposite condition, where the training samples are sufficiently large [79], and where the data samples may not correlate closely to each other. In this experiment, it was observed that the EDLS achieved improvement after the first iteration (i.e.,

$\bar{Pr}_{(t=1)} = 50.2\%$), because the retrieved data at t=0 usually has a low degree of correlation. Its performance, however, was reduced after two iterations as the retrieved samples became correlated more strongly. Using the same RBF widths, the OLS learning strategy was more stable and much better than the EDLS. This suggests that the EDLS may not be suitable for constructing the RBF network under this learning condition.

It was observed that the RBF centers critically influenced the performance of the RBF classifier, and that the RBF classifier constructed by matching all retrieved samples exactly to the RBF centers degraded the retrieval performance. The OLS algorithm was fairly successful at resolving this problem, by choosing the subset of the retrieved samples for the RBF centers. However, the OLS provided a less adequate RBF network, compared to the ARBFN. In the ARBFN learning, each available positive sample was considered as important; the centers were shifted by negative samples with the weighted norm parameters being updated during interactive cycles. The ARBFN also managed well with the small set of samples encountered. The proposed ARBFN network model is thus the most adequate model for the current application of interactive learning.

It is also interesting to see how the performance of the learning systems was influenced by the initial ranking results. The behavior of the systems was analyzed when only a few positive samples were found at the first iteration, i.e., the initial results were low, $\bar{Pr}_{(t=0)} \leq 50\%$. The precision results of 23 queries that had $\bar{Pr}_{(t=0)} \leq 50\%$ are shown in Fig. 3.4(a); note that these queries are a subset of all 35 test queries. For comparison, Fig. 3.4(b) shows the average precision of the other 12 queries at $\bar{Pr}_{(t=0)} > 50\%$. The queries in Fig. 3.4(a) may be regarded as "hard" ones, since they were retrieved at an average of only 37% precision at the first round. It is also observed that the learning systems seen in Fig. 3.4(a) took more iterations to converge to 90% precision, compared to those in Fig. 3.4(b). Interestingly, the ARBFN algorithm quickly converged to nearly 100% precision in only two iterations when $\bar{Pr}_{(t=0)} > 50\%$, and still worked well even when dealing with a small set of *positive* samples (Fig. 3.4(a)). In contrast, the OLS algorithm was quite slow to improve with the "hard" query set. The performance gap between the ARBFN and the OLS was striking in the case of 'hard' queries. Meanwhile, the EDLS showed almost no improvement if $Pr_{(t=0)} \leq 50\%$ (Fig. 3.4(a)), and fluctuated even after more iterations.

	Average precision (%), $\bar{Pr}_{(t)}$			
Method	t=0	t=1	t=2	t=3
ARBFN	44.82	80.72	90.36	92.50
EDLS	44.82	50.18	43.39	43.04
OLS	44.82	66.07	73.2143	76.61

Table 3.1. Average precision (%) as a function of iteration, $\bar{Pr}_{(t)}$, obtained by retrieving 35 queries, using Corel database.

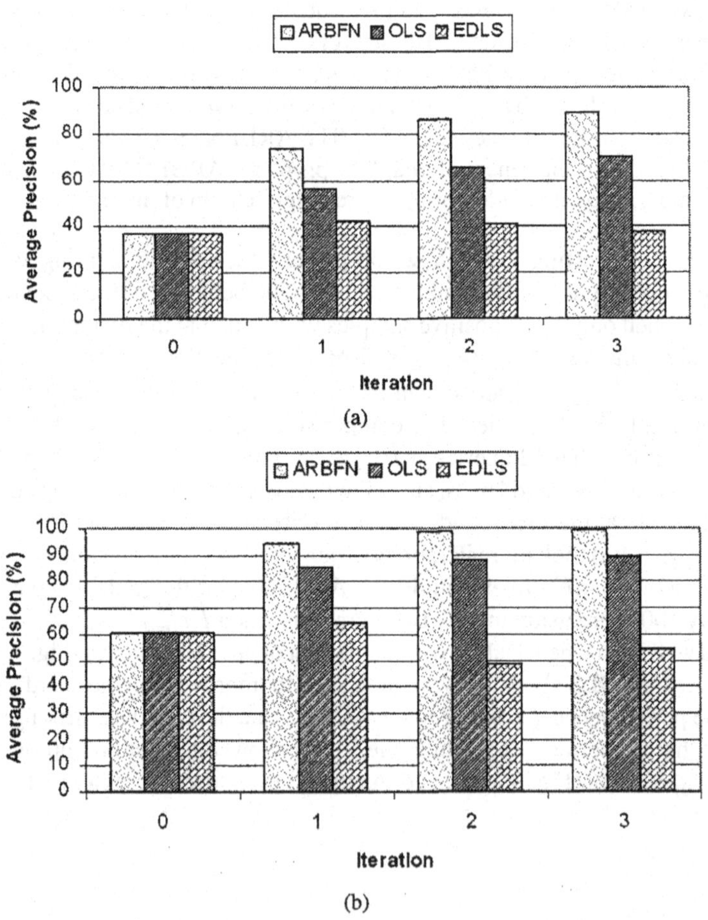

Figure 3.4. Retrieval results as a function of iteration; (a) average precision over 23 queries, initialed with less than 50%; (b) average precision over 12 queries, initialed with more than 50%.

A Comparison to the Global Model

The learning performance of the ARBFN learning strategy was next compared to the global model discussed in Chapter 2. The results (Table 3.1) were compared to the best results obtained in the previous experiments Section 2.6 (Table 2.5). The ARBFN achieved an improvement 1.6% higher than that of the single-RBF learning method. It was noted that, although the performance was just marginally better, the ARBFN was a more flexible learning method than the single-RBF, in the sense that it did not require parameter tuning. That is, as seen in the experiment in Chapter 2, for the single-RBF only the proper choice of parameters, $(\beta, \alpha_N, \alpha_R)$, is effective for learning capability and retrieval performance. This leads to difficulties in retrieval with unknown databases, since tuning of these parameters needs to be done for each database.

In the following experiments various learning methods were compared, using a new query set, which contained 59 images randomly selected from different categories (Fig. 3.5). The methods compared include ARBFN, single-RBF, OPT-RF [14], and MARS [8]. Two criteria were employed for performance measures: first, $Pr(10)$, $Pr(16)$, $Pr(25)$, and $Pr(50)$, i.e., the precisions after 10, 16, 25 and 50 images were retrieved; and second, a *precision* versus *recall* graph. Precision can be formally defined as:

$$Pr(N_c) = \frac{N_R}{N_c} \in [0, 1] \tag{3.37}$$

where N_R denotes the number of relevant image retrieved, and N_c denotes the number of retrieved images. Recall is defined as:

$$Re = \frac{N_R}{N_{TR}} \in [0, 1] \tag{3.38}$$

where N_{TR} denotes the total number of relevant images in the database. Using these measuring methods, the precision/recall graph represents the highest information content [80]. It should be noted that, for every measuring scheme, the relevance feedback process was done using the *top sixteen* retrieved images. This is to ensure that the learning methods remained valid for further practical application.

Table 3.2 summarizes the precision results averaged over all 59 queries, measured from the top 10, 16, 25, and 50 retrieved images[3]. It can be seen that the learning methods provided a significant improvement in each of the first three iterations. The ARBFN achieved the best precision results in all conditions, compared to the other methods discussed. At $N_c = 10$, ARBFN reached a near-perfect precision of 100% after three iterations. This means that *all* the top ten retrieved images were relevant. The results also show that, at $N_c = 16$, more than 14 relevant images were presented in the top 16 ranking set. The most important precision results are perhaps those after the first iteration, since users would likely provide only one round of relevance feedback. It was observed that the ARBFN provided a better improvement than the other methods for this requirement. Particularly, at $N_c = 25$, the ARBFN achieved 26%, while the single RBF was 21%, and the OPT-RF was 17%. The single-RBF needed one and two more rounds to match the first-round precision of the ARBFN at $N_c = 25$ and $N_c = 50$, respectively. In addition, the OPT-RF needed at least two more iterations to match the first-round precision of the ARBFN for $N_c \geq 16$.

Fig. 3.6(a), (b) and (c), illustrate the average precision versus recall figures after one, two, and three iterations, respectively. The behavior of the system without learning and the strong improvements with interactive learning can easily be seen. Note that part of these results are previously described in Table 3.2. In all cases, the precision at 100% recall drops close to 0. This fact indicates that it was not possible to retrieve all the relevant images in the database, which had been pre-classified by the Corel Professionals. It is observed from Fig. 3.6(a) that the ARBFN was superior to the single-RBF at the higher recall levels, while both provided similar precision at the lower recall levels. Also, the ARBFN achieved better improvements than the single-RBF by up to 8.6% , 7.3% and 6.5%, at one, two and three iterations, respectively.

Figs. 3.7-3.10 graphically illustrate the learning performances of the ARBFN compared with the single-RBF, for the queries, "CAR" and "SWIMMERS." Subjected to the first round of user's feedback, the results in Fig. 3.8 show that the ARBFN could retrieve more "CAR" than the single-RBF shown in Fig. 3.7. For the "SWIMMERS" query, it is assumed that the user may be interested in retrieving each relevant image in the pre-classified image set[4] shown in Fig. 3.11. It was observed that a single model cannot easily characterize the relevant images in this set. This is confirmed by the results of the single-RBF, shown in Fig. 3.9, where only the "exact matches" from the pre-classified relevant images were retrieved. In contrast, the results in Fig. 3.10 show that both the exact matches (retrieved in Fig. 3.9) and the other relevant models were retrieved using the ARBFN. This indicates the main advantage of utilizing the multiple-class modeling approach, which offers high performance classification power to the interactive learning system.

Figure 3.5. The 59 query images, obtained from different categories form the Corel database.

N_c	Method	Average precision (%), $Pr(N_c)$			
		t=0	t=1	t=2	t=3
10	ARBFN	55.93	+32.03	+42.03	+43.56
	Single-RBF	55.93	+27.97	+39.15	+42.03
	MARS	55.93	+17.12	+19.32	+19.66
	OPT-RF	55.93	+24.07	+30.51	+32.37
16	ARBFN	47.67	+30.83	+39.30	+41.21
	Single-RBF	47.67	+26.48	+34.64	+38.45
	MARS	47.67	+13.88	+16.00	+16.21
	OPT-RF	47.67	+20.97	+23.83	+25.00
25	ARBFN	39.93	+26.44	+30.44	+31.19
	Single-RBF	39.93	+21.36	+26.58	+28.14
	MARS	39.93	+11.46	+12.47	+12.07
	OPT-RF	39.93	+17.02	+19.73	+20.00
50	ARBFN	30.03	+19.08	+20.58	+20.75
	Single-RBF	30.03	+15.29	+17.76	+18.44
	MARS	30.03	+8.24	+8.31	+8.17
	OPT-RF	30.03	+11.86	+12.17	+12.51

Table 3.2. Average precisions, $\bar{P}r(N_c)$, compared at four settings of top matches, obtained by retrieving 59 queries, using the Corel Database. Interactive results are quoted relative to the $\bar{P}r$ observed with non-interactive retrieval. NOTE: At every N_c the top 16 samples were used for RF learning.

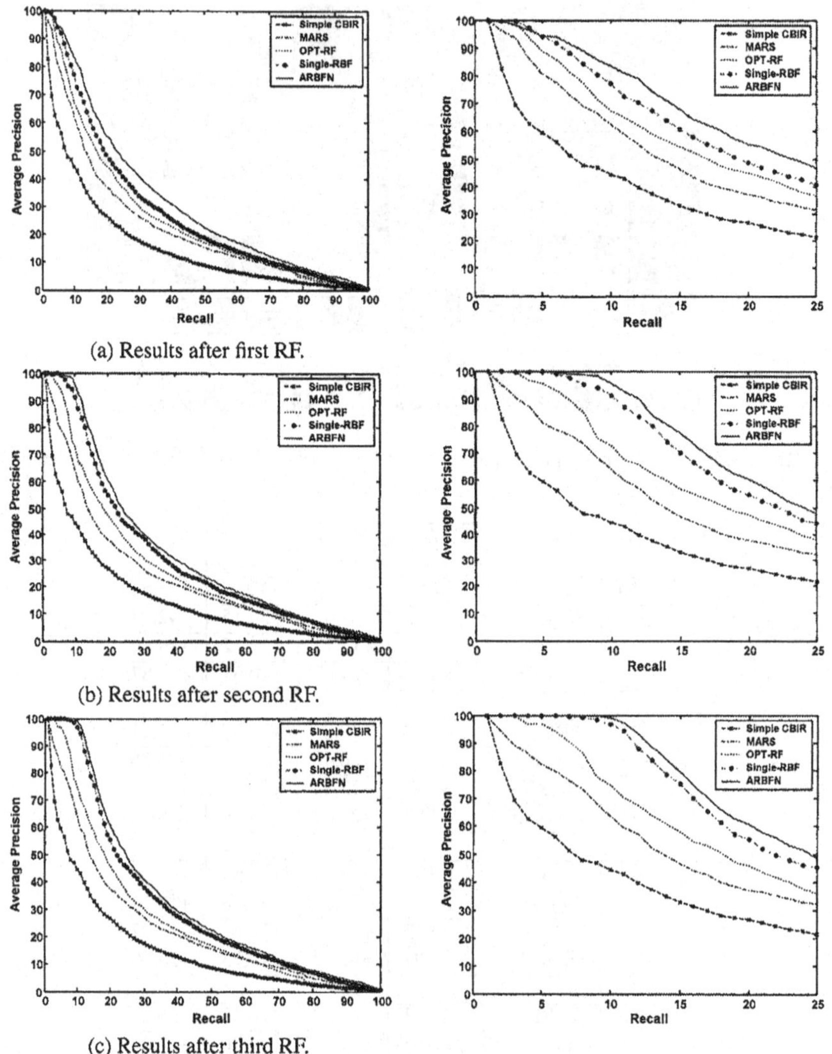

(a) Results after first RF.

(b) Results after second RF.

(c) Results after third RF.

Figure 3.6. Average precision versus recall figures, obtained by retrieving 59 queries, using Corel database. Figures on the right hand are the zoom version of the left hand figures. Note that, in each case, results obtained by simple-CBIR are fixed, and used as a benchmark for other interactive methods.

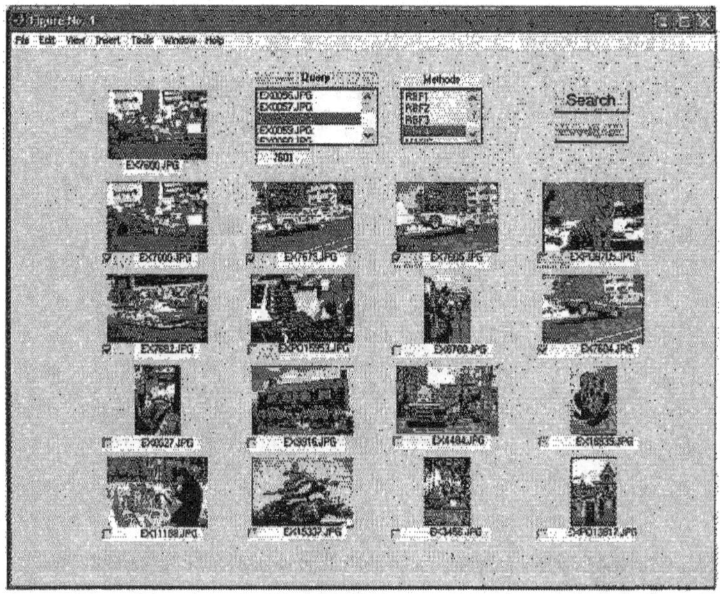

(a) Before learning (t=0), *Pr*=0.31.

(b) After learning (t=1), *Pr*=0.5.

Figure 3.7. Retrieval results of the query "CAR," obtained by single-RBF.

(a) Before learning (t=0), *Pr*=0.31.

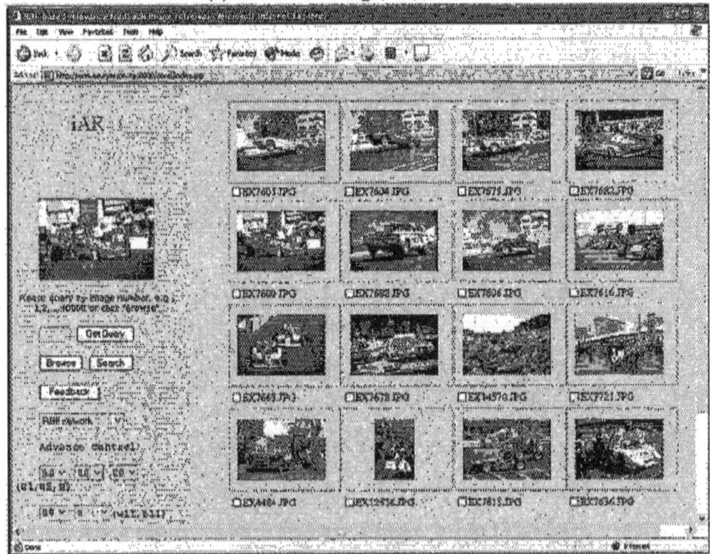

(b) After learning (t=1), *Pr*=0.75.

Figure 3.8. Retrieval results of the query "CAR," obtained by ARBFN.

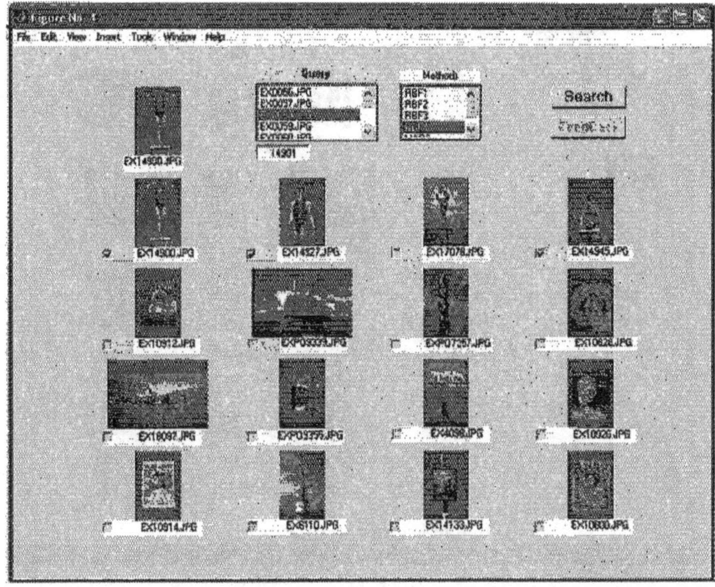

(a) Before learning (t=0), *Pr*=0.19.

(b) After learning (t=1), *Pr*=0.38.

Figure 3.9. Retrieval result of the query "SWIMMERS," obtained by single-RBF.

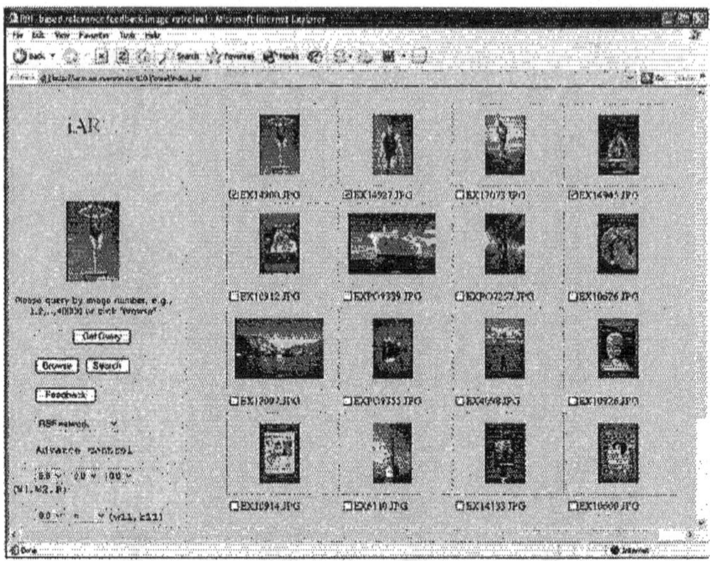

(a) Before learning (t=0), Pr=0.19.

(b) After learning (t=1), Pr=0.81.

Figure 3.10. Retrieval result of the query "SWIMMER," obtained by ARBFN.

7. Chapter Summary

A human-controlled interactive CBR system has been developed with local model network (LMN) architecture which directly assimilates the essential

Figure 3.11. Pre-classified images group representing "SWIMMERS," obtained by the ground truth from the Corel professionals. Within this group, there are a total of 100 similar images, out of 40,000 image in the database.

characteristics of human specified features through a learning process. The specific architecture of the network divides the training features into sub-classes in such a way as to reflect the different preferences of users regarding image similarity. Conventional image retrieval algorithms using relevance feedback (RF), mainly focus on a global model in learning parameters for characterization of image similarity; that is, they use a single adjustable query model and its associated mapping function, where data distribution is assumed to be Gaussian. The current system implicitly represents this distribution in the form of a mix of Gaussian weights, which are updated during the training process to cope with nonlinear distributions, to fully exploit sub-class information. The use of this local characterization of image relevancy, coupled with the use of nonlinear analysis, offers a better evaluation of image similarity, suggesting how clustering techniques can be used to model the local context as defined by a current query session. This approach requires very little parameter tuning.

It is also important to note that, in contrast to other learning methods used to identify RBF network models, the proposed learning strategy takes into account the nature of relevance feedback problems, in that the training samples are of high correlation and limited in size. This nature makes the current application differ from the general function approximation. The proposed learning strategy shows the adaptation of the RBF network architecture and has been successfully applied to interactive image retrieval with promising results. Experiments have shown that the adaptive RBF network is capable of generalizing the information from minimal training data. Compared to global RF learning methods, the experimental results also show that it can achieve higher retrieval accuracy and quicker convergence. One further attractive feature of the adaptive RBF network approach is its capability to capture sub-classes of information of the specified relevant images, rather than only nearly exact matches.

Notes

1 The MATLAB function `newrb` was employed.

2 In order to compare with the first function, the MATLAB function `newrbe` was also employed.

3 It was assumed that the average number of images that most interests the user ranges between 10 and 50.

4 the ground truth classes were used to obtain the relevant images in Fig. 3.11.

Chapter 4

AUTOMATIC RELEVANCE FEEDBACK

1. Introduction

This chapter discusses machine controlled interactive content-based retrieval (MCI-CBR) techniques using a self-learning architecture. Specifically, the self-organizing tree map (SOTM) is adopted to minimize user participation and to automate the relevance feedback learning methods discussed in the previous chapters. The main advantage of the MCI-CBR approach over human-controlled interactive content-based retrieval (HCI-CBR) approaches, is in the automation of the process of interaction. Minimizing user participation provides a more user-friendly environment, and avoids errors caused by excessive human involvement. In HCI-CBR systems, users are required to bring the systems up to the desired level of performance. These systems depend on a large number of query submissions (in the form of user relevance feedback samples) to improve retrieval capability and impose a great responsibility on the user. Although existing interactive learning methods can effectively learn with a small set of samples, it is still necessary to reduce the user workload. Moreover, reduced feedback also decreases the number of transmitted images or videos from the service provider, reducing the required transmission bandwidths, an important requirement for Internet-based retrieval applications.

The main interest in this chapter is to integrate the MCI-CBR system with compressed domain processing in support of remote access to digital libraries over the Internet. In this context, it is assumed that users are most interested in accessing many sources of information distributed and shared in the image databases across the Internet. Although images and videos stored in those databases may be indexed and organized in various ways, data is usually stored in a compressed format, employing a number of existing multimedia compression techniques. Given this, it is therefore appropriate to introduce compressed domain indexing and retrieval. This enables the realization of direct access to compressed data without employing any pre-organized database indexing.

The MCI-CBR learning system is implemented by a novel neural network, the self-organizing tree map (SOTM) [35, 36], which has been proven to be effective in image processing applications such as unsupervised image segmentations [94], denoising, and compression [35, 36]. Unlike HCI-CBR systems, where the user's direct input is required in the execution of the RF algorithms, the SOTM estimate is now adopted to *guide* the adaptation of the RF parameters. As a result, instead of imposing a greater responsibility on the user, independent learning can be integrated to improve retrieval accuracy. This makes it possible to obtain either a fully automatic or a semiautomatic RF system suitable for practical application. Consequently, the interactive CBR system of Fig. 4.1(a) is generalized to include a self-learning component, as shown in Fig. 4.1(b). The interaction and relevance feedback modules are implemented in the form of specialized neural networks. In these fully automatic models, the learning capability associated with the networks and their ability to perform general function approximations, offers improved flexibility in modeling the user's preferences according to the submitted query.

The Architecture of the MCI-CBR System with Compressed Domain Processing

The system architecture, shown in Fig. 4.2(a)-(b), is composed of a compressed domain search unit and a self-organizing adaptation network. In the initial stage, the search unit utilizes compressed-domain visual descriptors to access the compressed databases, and the similarity is simultaneously computed. In the following adaptation stage, the self-organizing network is applied to the retrieved samples to make a decision for advising on the relevance feedback module and for improving retrieval results. The system can either run in fully automatic mode (Fig. 4.2(a)) or can be incorporated into the user's interactions to learn further different user subjectivities (Fig. 4.2(b)). During automatic learning, there is no transmission of the sample files from the server side to a user client, as only the improved retrieval set will be delivered to the user. Thus, the required transmission bandwidth can be reduced or eliminated during the retrieval process.

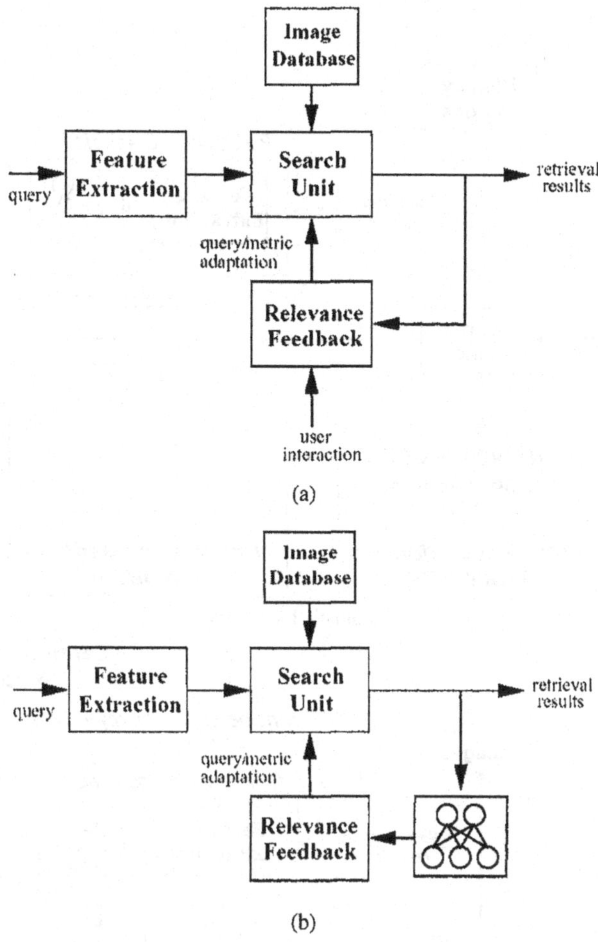

Figure 4.1. (a) HCI-CBR system (b) MCI-CBR system.

2. Automatic Interaction by Self-Organizing Neural Network

The user's judgment of image relevancy is closely related to the classification problems associated with the supervised learning of artificial neural networks. Here, the user's information is used to enhance the network's classification power. Contrary to this supervised learning, a special class of neural net-

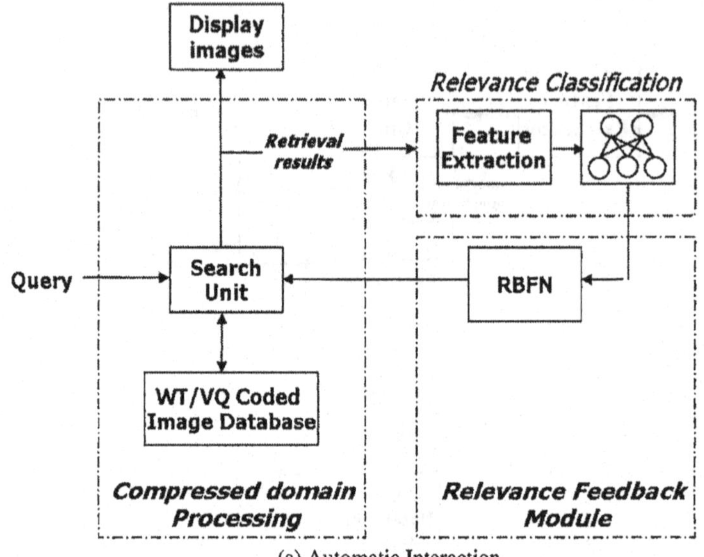

(a) Automatic Interaction.

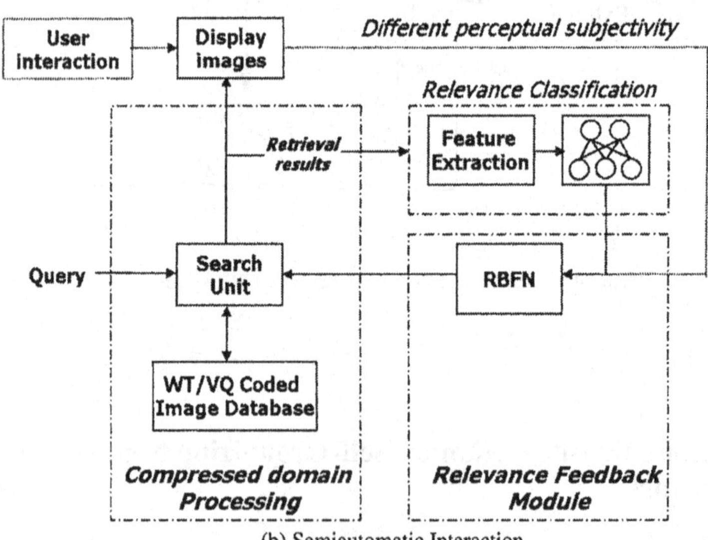

(b) Semiautomatic Interaction.

Figure 4.2. MCI-CBR architecture with compressed domain processing.

works known as a self-organizing map (SOM), approaches pattern classification problems based on competitive learning *without* an external teacher [85]. This unsupervised learning paradigm is employed here for the classification of relevance in interactive retrieval. In particular, the self-organizing tree map (SOTM) [35, 36] is adopted in this work.

The SOTM is a tree structured SOM [35, 36]. It is a hierarchical approach for the purpose of data clustering. One chief advantage of the SOTM over classical unsupervised learning methods, such as SOM and k-mean algorithms, is that it has the ability to work well with sparsely distributed data [34]. For this reason, the SOTM was chosen for this work for automatic relevance identification. The source of such sparse data in the current application is the high dimension of the data space (e.g., image feature vectors). There is also a limit to the size of training data to which we can afford to label individual items of the retrieved data, for subsequent relevance feedback learning.

Figures 4.3(a)-(b) show a comparison of the clustering performances between the SOTM and the SOM on a two-dimensional feature space [34]. In these figures, the input vectors are uniformly distributed within many rectangular squares. The SOTM's clustering performance in Fig. 4.3(a) shows that there are no neurons lying in a zero-density area. In contrast, although the SOM's topology exhibits the distribution of the structured input vectors, it also introduces several false representations outside of the distribution of the input space, as shown in Fig. 4.3(b). As indicated by this example, the SOTM realises the learning not only of the weights of the connections, but also of the suitable structure of the network. This property is necessary for a high-dimensional input space with a sparse data structure, in order to classify image relevancy in the current application.

Feature Space for Relevance Identification

Having selected the learning architecture, we focus on a feature space which can effectively distinguish between images of relevance and irrelevance. The general guideline for selecting the feature space in relevance identification is as follows: in automatic retrieval, two feature spaces are required: $\mathcal{F}_C$ and $\mathcal{F}_S$. $\mathcal{F}_C$ is for retrieval, and $\mathcal{F}_S$ is for relevance identification. This reflects the fact that we have different requirements for $\mathcal{F}_C$ and $\mathcal{F}_S$. Since retrieval is performed on huge volumes of data, the primary requirement for $\mathcal{F}_C$ is high speed and reasonable accuracy. Normally, compressed domain features are used for this requirement, as images are stored and transmitted in the compressed domain. On the other hand, relevance identification requires high accuracy,

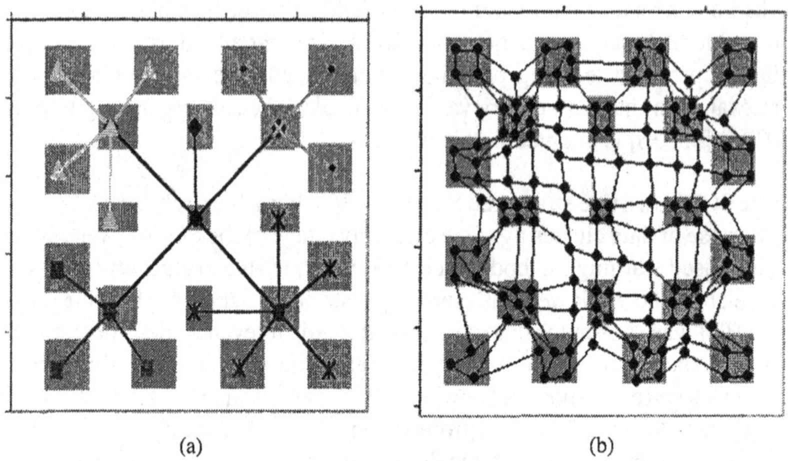

Figure 4.3. Self-organizing data clustering: (a) performed by SOTM, no nodes converge to join areas of zero data density; (b) performed by SOM, nodes converge to areas of zero data density.

since we want the process to simulate human performance. Hence the features in $\mathcal{F}_S$ should be of high quality. While use of the feature space $\mathcal{F}_S$ may substantially increase the computational complexity required per image, the added cost is minor, since only a small number of retrieved images undergo such computation. Moreover, the sophisticated data structure of feature space $\mathcal{F}_S$ will lead to better identification of image relevance.

The notions of cross-correlation between the two feature spaces are formally described below:

Let $\mathcal{T}_I$ denote the set of retrieved images, $\mathcal{I}_i, i = 1, 2, ..., N_{RT}$, *which is obtained from the database using the feature space* $\mathcal{F}_C$, *employing the "nearest neighbor search". Each retrieved image is then projected onto the feature space* $\mathcal{F}_S$ *by* $\mathcal{S} : \mathcal{R}^{spatial} \to \mathcal{F}_S$, *which produces a feature vector* $\mathbf{v}_i = \mathcal{S}(\mathcal{I}_i)$.

As a result of the projection, the set of retrieved images is employed as a training data set, $\mathcal{T}_{(t)} = \{\mathbf{v}_1, \mathbf{v}_2, ..., \mathbf{v}_{N_{RT}}\}$, which will be used for relevance identification by the SOTM algorithm. The algorithms used to obtain the descriptors in the spaces $\mathcal{F}_C$ and $\mathcal{F}_S$ are explained in Section 3.

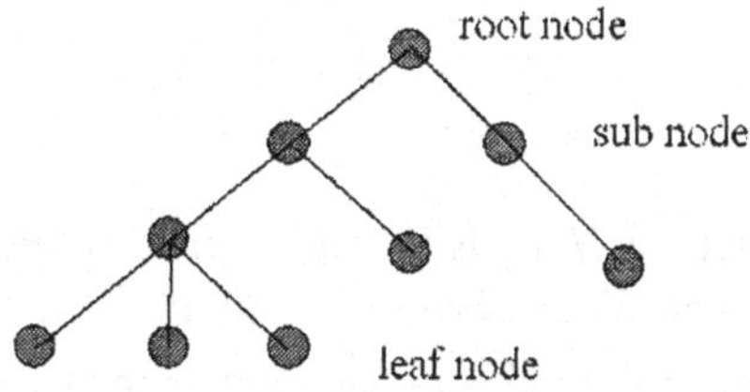

Figure 4.4. The tree hierarchy structure of SOTM.

SOTM Algorithm

The SOTM is a competitive neural network which uses a multi-layer tree structure to organize the neurons, as shown in Fig. 4.4. Following the statistics of the input pattern, different neurons are constructed to capture the local context of the pattern space and mapped into the tree structure. In other words, the map attempts to represent all available samples with optimal accuracy, using an adaptively defined set of models. At the same time, the models become ordered on the tree map so that similar models are close to each other and dissimilar models far from each other. Consequently, the different descriptions of relevant and non-relevant images can generate their own centres on the tree map, with the application of SOTM to the relevance classification.

In order to construct a suitable map, the SOTM offers two levels of adaptation: weight and structure. The weight adaptation is the process of adjusting the weight vector of the "winning" neurons. Structure adaptation is the process of adjusting the structure of the network by changing the number of neurons and the structure relationships between them. Given a training data set $\mathcal{T}_{(t)} = \{\mathbf{v}_1, \mathbf{v}_2, ..., \mathbf{v}_{N_{RT}}\}$, the adaptation map using the SOTM algorithm is summarized as follows [36]:

Step 1. *Initialization.* Choose the root node $\{\mathbf{w}_j\}_{j=1}^{l}$ with a randomly selected training vector.

Step 2. *Similarity matching.* Randomly select a new data point $\mathbf{v} = [v_1, ..., v_i, ..., v_P]^T$, and compute the Euclidean distance, d_j to node $\mathbf{w}_j, j = 1, 2, ..., l$:

$$d_j = \sqrt{\sum_{i=1}^{P} (v_i(n) - w_{ji}(n))^2} \quad (j = 1, ..., l) \tag{4.1}$$

Step 3. *Updating.* Select the winning node, j^*, with minimum d_j, $\forall j$:

$$d_{j^*} = \min_j d_j(\mathbf{v}, \mathbf{w}_j) \tag{4.2}$$

If $d_{j^*} \leq H(n)$, where $H(n)$ is the hierarchy function used to control the levels of the tree (and $H(n)$ decreases with time),

Then assign $\mathbf{v}$ to the j^*-th cluster, and update the weight vector according to the reinforced learning rule:

$$\mathbf{w}(n+1) = \mathbf{w}(n) - \alpha(n)[\mathbf{v}(n) - \mathbf{w}(n)], \tag{4.3}$$

where $\alpha(n)$ is the learning rate, which decreases with time, $0 < \alpha(n) < 1$,

Else form a new subnode starting with $\mathbf{v}$.

Step 4. *Check terminate condition.* Exit if one of the following three conditions is fulfilled.

- The specified number of iterations is reached.
- The specified number of clusters is reached.
- No significant change to the tree map has occurred.

Otherwise repeat by going back to Step 2.

Scanning the input data in a random manner is essential for convergence in the algorithm [34, 94]. Consider that the data is scanned in such a way that the ith component of feature vector $\mathbf{v}$ is monotonically increasing, then the ith component of the weight vector $\mathbf{w}_j$ will also be monotonically increasing, $\forall j = 1, 2, ..., l$ [according to step 3]. Thus *all of the nodes* will have a monotonically increasing component in the ith position of the weight vector.

The hierarchy control function $H(n)$ controls the levels of the tree. It is initialled with a large value, and decreases with time. With the decrease of $H(n)$, a subnode comes out to form a new branch, and recursively progresses until it reaches the leaf node. This process produces the entire tree structure, which preserves topological relations from the root node to the leaf nodes.

Subnode Selection

After convergence, the SOTM algorithm produces new node vectors $\breve{\mathbf{w}}_j, j = 1, 2, ..., L$. Let $\mathbf{v}_{query} \in \mathcal{F}_S$ be the feature vector associated with a given query image in the current retrieval session. By going along the tree structure, we can find nodes close to the given query. We calculate a set of distances, $d_j, j = 1, 2, ..., L$, between the query vector and the node vectors:

$$d_j = \left((\breve{\mathbf{w}}_j - \mathbf{v}_{query})^T (\breve{\mathbf{w}}_j - \mathbf{v}_{query}) \right)^{\frac{1}{2}}. \tag{4.4}$$

Based on the calculation result, the nodes:

$$\{\breve{\mathbf{w}}_{j^*} |\, j^* = 1, ..., L', \quad L' < L\} \tag{4.5}$$

that are closest to the query vector are then defined to be relevant, so that any input vector partitioned in these clusters is assessed as relevant. Ideally, the number of nodes L' chosen will optimize the classification performance.

Corresponding to these relevant nodes, we define regions $R_{j^*} \subset \mathcal{F}_S$ to associate with the relevant nodes $\breve{\mathbf{w}}_{j^*}$. These regions are then used for classification of the retrieved images. A class label, $y \in \{1, 0\}$, assigned to a given retrieved image, $\mathcal{I}_i, i \in \{1, ..., N_{RT}\}$ is then obtained:

$$y_i = \begin{cases} 1 & \mathbf{v}_i \in R_{j^*}, \forall j^* \\ 0 & \text{otherwise} \end{cases} \tag{4.6}$$

The network output is formally defined as $(\mathcal{I}_i, y_i)$, $i = 1, 2, ..., N_{RT}$, where y_i is the class label obtained for the retrieved image $\mathcal{I}_i$. This result will guide the adaptation of relevance feedback modules in the interactive retrieval sessions.

Automatic and Semi-automatic Interaction Procedures

SOTM has been used to guide the adaptation of RF-based learning methods, in both automatic and semi-automatic modes. The learning methods include: (1) single-RBF method described in Chapter 2, (2) Adaptive-RBF Network described in Chapter 3, and (3) relevance feedback method (RFM) [9, 25]. The retrieval procedures using machine interactions are summarized as follows:

Step 1. *Self-organizing adaptation*: For a given query image, the most similar images, $\mathcal{I}_i$, $i = 1, ..., N_{RT}$, are obtained by using a Euclidean distance measure based on the pre-defined vectors in the feature space $\mathcal{F}_C$. The retrieved images are then input to the SOTM algorithm to identify relevance, employing a new feature space $\mathcal{F}_S$. This constructs the training set $(\mathcal{I}_i, y_i)$, $y_i \in \{0, 1\}$, $i = 1, 2, ..., N_{RT}$.

Step 2. *RF adaptation*: A new search operation is performed on the database using Single-RBF, ARBFN, or RFM, employing the training data, $(\mathcal{I}_i, y_i)$, $i =$

$1, 2, ..., N_{RT}$, in the feature space $\mathcal{F}_C$. The process may either be repeated by inputting the new retrieved image set to the SOTM in Step 1, or stopped. This process may also be switched to run in semiautomatic mode by incorporating user interactions.

Apart from the automatic adaptation, an important advantage to using a self-organizing network is the flexibility in choosing the number of training samples, N_{RT}. It allows for learning with a large number of training samples, which is usually not possible in HCI-CBR. As a result, interactive learning methods, (e.g., single-RBF and ARBFN), increase their learning capabilities with a large sample number.

The following two sections describe the feature descriptors used for retrieval and relevance identification in the interaction process.

3. Descriptors in JPEG and Wavelet Transform (WT) Domains

Although the gap in human perception and visual descriptors could be more significant in the transform domain than in the spatial domain, compressed-domain descriptors are widely used in image retrieval [38, 104–107, 109]. The main reason for this is that feature extraction algorithms can be applied to image and video databases without full decompression. This provides fast and efficient tools that are appropriate for real-time applications. The following subsections investigate the characteristics of discrete cosine transform (DCT) and discrete wavelet transform (DWT) domains, and construct applicable schemes for extracting compressed-domain descriptors.

DCT Domain Visual Descriptor

Largely due to the "energy packing" property, DCT has been widely used in JPEG and many other popular image and video compression standards. When a typical 8×8 block of data undergoes DCT transformation, most of the significant coefficients are concentrated in the upper-left (low frequency) region of the transform block, thus allowing the storage and transmission of a small number of coefficients.

In this study, compressed-domain descriptors were employed using the energy histogram of the DCT coefficients originally proposed by Lay [38]. An energy histogram of DCT coefficients is obtained by counting the number of

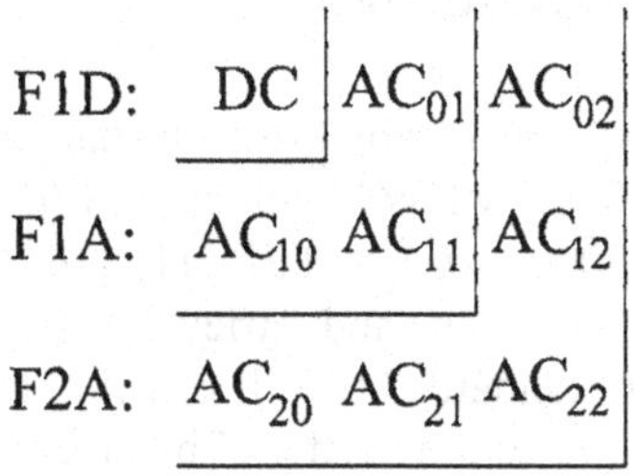

F1D: DC AC_{01} AC_{02}

F1A: AC_{10} AC_{11} AC_{12}

F2A: AC_{20} AC_{21} AC_{22}

Figure 4.5. The three coefficient groups in the DCT domain.

times a particular coefficient value appears in an 8×8 block. Formally, the value of the histogram in the mth bin can be written as:

$$h_c[m] = \sum_{u=0}^{7} \sum_{v=0}^{7} I(Q(F[u, v]) = m) \tag{4.7}$$

where $Q(F[u, v])$ denotes the value of the coefficient at the location (u, v), and m is the index of the current histogram bin. The function $I(\cdot)$ is equal to 1 if the argument is true, and 0 otherwise.

For a chrominance DCT block, the DC coefficient is proportional to the average of the chrominance values in the block. As a result, the histogram of DC coefficients can be used as an approximation of the color histogram of the original image. On the other hand, the histogram of the AC coefficients can be used to characterize the frequency composition of the image.

In this work, the 9 DCT coefficients in the upper left corner of the block are partitioned into three sets as follows (Fig. 4.5):

$$\begin{aligned} F1D &= \{DC\} \\ F1A &= \{AC_{10}, AC_{11}, AC_{01}\} \\ F2A &= \{AC_{20}, AC_{21}, AC_{22}, AC_{12}, AC_{21}\} \end{aligned}$$

For the experiments in Section 5, the energy histogram features are based on the coefficients in two of these collections:

$$F = F1D \cup F1A = \{DC, AC_{01}, AC_{10}, AC_{11}\}. \tag{4.8}$$

Separate energy histograms are constructed for the DC and AC coefficients of each of the colour channels, and 30 bins are used for each histogram.

DWT Domain Visual Descriptor

Many recent image/multimedia compression standards employ state-of-the-art compression technologies by DWT, including MPEG-4 Visual Texture Coding (VTC) [101], JPEG 2000 [102], and SPIHT[1] [103]. This trend has caused much of the latest work on indexing and retrieval to focus on algorithms that are compatible with the new standards [104, 106, 108]. In JPEG2000, image descriptors can be extracted from the header of bitstream packets [105], or based on region [109].

In the wavelet-baseline coders, the difference among the wavelet coders is only in the process of encoding wavelet coefficients. Specifically, in JPEG2000, the subband samples are partitioned into small blocks of samples, called "code-blocks." Each code-block is encoded independently. In SPIHT, the wavelet coefficients are coded based on the self-similarity across scales of the wavelet transform using the tree-based organization of the coefficients. For this reason, in order to make the feature extraction algorithms compatible with all of the coders, it it best to extract the descriptors directly after the decomposition process, before encoding the DWT coefficients.

Fundamental Statistic Descriptor (FSD)

It is well known that the histogram of wavelet 'detail' coefficients can be modelled by the generalized Gaussian density (GGD) function [106, 110, 111], which is defined as

$$p(x; \alpha, \beta) = \frac{\beta}{2\alpha\Gamma(1/\beta)} e^{-(|x|/\alpha)^{\beta}}, \tag{4.9}$$

where α models the width of the probability density function, β is the shape parameter, and $\Gamma(\cdot)$ is the Gamma function, i.e., $\Gamma(z) = \int_0^\infty e^{-t} t^{z-1} dt$, $z > 0$. Do *et al* [106] used the parameters α and β, estimated from each subband, as the compressed-domain descriptor. Similarly, the variance of coefficients within the code-blocks is used to generate the descriptor for JPEG2000 coded images [104].

In the experiments, a simple solution is obtained by computing the mean and the standard deviation of wavelet coefficients in each subband [95]. Let the wavelet coefficients at the ith subband be described by $x_{i,1}, x_{i,2}, ..., x_{i,P}$ for three level decompositions, i.e., $i = 10$. The two parameters are then computed for each subband, $m^{(i)}, \alpha^{(i)}$ where $m^{(i)} = \bar{x}_i$ and $\alpha^{(i)} = var\{x_{i,1}, x_{i,2}, ..., x_{i,P}\}$.

These descriptors are referred to as *fundamental statistic descriptors* (FSD), which can be applied to all types of wavelet-based coding standards.

Feature Extraction Algorithm for Wavelet Transform/Vector Quantization (WT/VQ) Coders

DWT (Discrete wavelet transform), In order to further reduce the decompression process, and increase wavelet coder specificity, a feature extraction algorithm [15] is applied. The hybrid WT/VQ coders have been proven to achieve high compression while maintaining good visual quality [28, 110]. A two-level wavelet decomposition scheme, with a 15-coefficient biorthogonal filter, is adopted for the analysis of images. The decomposition scheme produces three subbands at resolution level 1, and four subbands at resolution level 2. To capture the orientational characteristics provided by the wavelet-based subband decomposition, a codebook is designed for each subband. This results in a multiresolution codebook that consists of subcodebooks for each resolution level and preferential direction. Each of these subcodebooks is generated using the Linde-Buzo-Gray (LBG) algorithm, as well as the bit allocation method, to minimize overall distortion. The codebook design divides a subband belonging to different images in $m \times m$ square blocks and the resulting vectors are used as the training vectors. Separate codebooks are trained for each resolution-orientation subband.

For the encoding of an input image at a total bit rate of 1 b/pixel, the bit assignment is organized as follows: resolution 1 (diagonal orientation) is discarded; Resolution 1 (horizontal and vertical orientations) and resolution 2 (diagonal orientation) are coded using 256-vector codebooks ($m = 4$) resulting in a 0.5-b/pixel rate, whereas resolution 2 (horizontal and vertical orientations) is coded at a 2-b/pixel rate using 256-vector codebooks ($m = 2$); and, finally, the lowest resolution is coded by scalar quantization at 8-b/pixel.

The outcome of the coding process, referred to as coding labels, is used to constitute a feature vector via the computation of the labels histograms. Each subband is characterized by one histogram and represents the original image at a different resolution; thus, the resulting histograms are known as multiresolution histogram indexing (MHI) [15]. MHI features make use of the fact that the usage of codewords in the subcodebook reflects the content of the encoded input subimage. To address the invariant issues of the illumination level, only five subbands containing the wavelet 'detail' coefficients are concatenated to obtain the MHI features:

$$MHI = [h_1^{(1)} h_2^{(1)} \cdots h_{256}^{(1)} h_1^{(2)} h_2^{(2)} \cdots h_{256}^{(5)}]^T \qquad (4.10)$$

where $h_i^{(j)}$ denotes the number of times codeword i in subcodebook j is used for encoding an input image.

In order to reduce the dimension of the MHI vector, a feature selection method is applied. There are many techniques which have been proposed for feature selection, such as principle components analysis and sequential selection methods [96, 97]. In this work, the consistency-based feature selection method proposed in [100] is adopted, because of its simplicity and effectiveness. The idea is to select a subset of features from the original feature set, which is compromised of tradeoff between retrieval accuracy and computational complexity.

Let $F = \{f_1, ..., f_N\}$ denote the original feature set with N features. The purpose is to generate $F' = \{f'_1, ..., f'_M\} \subset F$ that contains the index of the relevant feature subset with $M < N$. To evaluate relevancy of each feature variable, a consistency measure is defined as [100]:

$$\rho = \frac{\text{mean interclass distance}}{\text{mean intraclass distance}} \tag{4.11}$$

For classification purposes, we need the intra-class as compact as possible, while the inter-class should be as separate as possible. This leads to a measure of consistency such that a better feature gives a greater value of ρ.

Fig. 4.6(a) plots the consistency measure, as a function of the index of the MHI feature variables, obtained from the experiments. The MHI features have been used to characterize texture images in the Brodatz database[2]. The results show that the degree of consistency is different between the features. Based on this measure, 13 subsets of MHI features have been generated and tested their retrieval effectiveness. Fig. 4.6(b) plots average retrieval rates (AVR), which were obtained by using all of the subsets, based on the Euclidean metric. It can be seen that subsets that contain more than 600 feature variables are good for approximating the original feature set. These subsets provide more than 62% AVR results, which indicates the tradeoff between the retrieval accuracy and the computational complexity.

4. Visual Descriptor for Relevance Classification

For relevance classification, it is required that a visual descriptor should be of high quality, in the sense that it can approximate a relevance evaluation as performed by human users. A sophisticated visual descriptor extracted by spatial domain techniques, objects, and region-based segmentations, would be appropriate for this requirement. Although, in general, these feature extraction

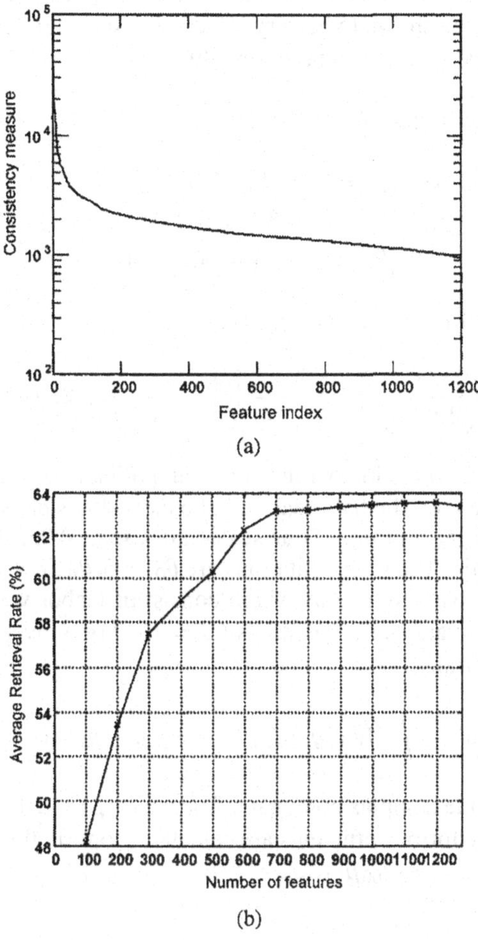

Figure 4.6. (a) Consistency measure as a function of index feature variables; (b) AVR results obtained by using the Euclidean metric, employing 13 different subsets of features, using 116 queries on the Brodatz database.

techniques are associated with high computational complexity, they are sufficiently quick for a small set of images presented in the interactive session, as compared to the whole database.

This work has chosen a Gabor wavelet transform technique [24, 88] as a visual descriptor for relevance classification. This technique is considered a powerful low-level descriptor for image search and retrieval applications. It has been used by the MPEG-7 to provide a quantitative characterization of homogeneous texture regions for similarity retrieval [98]. The Gabor wavelet method has also been shown to be very effective in such applications as texture classification and segmentation tasks [99, 89].

A bank of filters is employed, defined by:

$$g_{mn}(x,y) = a^{-m} g(x',y') \tag{4.12}$$
$$x' = a^{-m}(x\cos\theta_n + y\sin\theta_n) \tag{4.13}$$
$$y' = a^{-m}(-x\sin\theta_n + y\cos\theta_n), \tag{4.14}$$

which are derived from the 2-D Gabor function:

$$g(x,y) = \left(\frac{1}{2\pi\sigma_x\sigma_y}\right)\exp\left[-\frac{1}{2}\left(\frac{x^2}{\sigma_x^2}+\frac{y^2}{\sigma_y^2}\right)+2\pi jWx\right], \tag{4.15}$$

where $a > 1$, m and n are index filter scales and orientations, and W gives the modulation frequency. The half peak radial bandwidth is chosen to be octave, which determines σ_x and σ_y. Four scales are used: 0.05, 0.1, 0.2 and 0.4 cycles/image-width. The six orientations are $\theta_0 = 0, \theta_{n+1} = \theta_n + \pi/6$. This results in a set of basic functions which consist of Gabor wavelets spanning four scales ($S = 4$) and six orientations ($O = 6$). For a given image $\mathcal{I}(x,y)$, its Gabor wavelet transform is defined by:

$$W_{mn}(x,y) = \int \mathcal{I}(x_1,y_1) g^*_{mn}(x-x_1, y-y_1)dx_1dy_1 \tag{4.16}$$

where $*$ indicates the complex conjugate. The mean (μ) and the standard deviation (σ) of the transform coefficients are calculated to form the 48-dimensional feature vectors, $\mathbf{x} = [\mu_{00}\sigma_{00}\mu_{01}\ldots\mu_{(S-1)(O-1)}\sigma_{(S-1)(O-1)}]^T$.

5. Retrieval of JPEG and Texture Images in Compressed Domain

The following experiments are designed to compare the performances of four methods: non-interactive CBIR, user interaction retrieval, automatic retrieval, and semiautomatic retrieval. The experimental results are obtained from two image databases: the Brodatz database (DB1), which contains 1856 texture images; and DB2, which is distributed by Media Graphic Inc. [39], consisting of nearly 4700 JPEG color images covering a wide range of real-life photos, with completely open domain (see Fig. 4.7). These are typical, medium-size databases that are potentially accessible remotely through Internet environments, without pre-organization of the stored images [38].

The feature extraction algorithms described in Section 3 were applied to the compressed images. For the DB1 test set, the visual descriptor used is texture. Both FSD and MHI feature representations were tested, for the characterization of the wavelet-compressed images. The FSD descriptor was obtained before the invert-DWT process of the wavelet-baseline coder, whereas the MHI descriptor was obtained before VQ-decoding of the WT/VQ coder. For the DB2 test set, the visual descriptors used are the energy histograms which were extracted directly from the compressed JPEG images after the entropy decoding process.[3]

In the simulation study, 116 images from all texture classes were used as the query images for testing DB1. The performance was measured by the average retrieval rate, AVR (%), obtained from the top 16 retrievals. As for the DB2 database, 30 different query images were used for retrievals [cf. Fig. 4.7], and the performance was measured by the average precision. The relevance judgments were conducted using two criteria: (1) the ground truth; and (2) the subjectivity of the individual user. For the first criterion, the retrieved images were judged to be relevant if they were in the same class as the query. For the second criterion, the retrieved images were judged as relevant to the perception of the individual user.

Noninteractive Retrieval Versus Automatic Interactive Retrieval

Table 4.1 provides the numerical results illustrating the performances of the automatic interaction under different learning conditions. The results were obtained from the DB1 database employing FSD and MHI descriptors. In all cases, relevance judgment was based on the ground truth. The ARBFN method, the single-RBF method, and the RFM were tested. For each learning method, $N_{RT} = 20$ from the top ranked retrievals were utilized as training data. These samples were then input into the SOTM algorithm for the identification of relevance. The output of the unsupervised network was in turn used as the supervisor for a learning method to update learning parameters and to obtain a new set of retrievals. The learning procedure was allowed to continue for four iterations. The first column of AVR results corresponds to the initial results obtained by noninteractive retrieval methods. The remaining columns are obtained with the automatic interaction methods.

Evidently, the use of automatic learning techniques resulted in a significant improvement in retrieval performance over that of the simple CBIR technique. For the automatic ARBFN, 18.4% AVR improvement was achieved through four interactions, and 13% form automatic single-RBF. These retrievals used the

Figure 4.7. Example of images from DB2, each of which was used as a query.

FSD descriptor. The results for each learning method, with the MHI descriptor, show the same general trend. Maximum improvement was 13% with automatic ARBFN, and 12% with automatic RFM.

The observations reported in Chapters 2 and 3 are confirmed. The results in Table 4.1 show that ARBFN gave the best retrieval performance, compared to the other leaning methods, regardless of the descriptor types. In contrast, for single-RBF and RFM, the values for α_N and $(\alpha, \gamma, \varepsilon)$ significantly affected the results.

Based on FSD descriptor						
Method	0 Iter.	1 Iter.	2 Iter.	3 Iter.	4 Iter.	Parameters
ARBFN	58.78	69.02	72.85	76.24	77.21	-
Single-RBF	58.78	66.32	68.80	70.04	71.87	$\alpha_N = 0.1$
RFM	53.88	57.11	59.00	60.45	60.78	$\alpha = 1, \gamma = 1, \varepsilon = 0.5$
Based on MHI descriptor						
ARBFN	63.42	71.66	75.22	75.86	76.51	-
Single-RBF	63.42	70.31	72.74	73.11	73.06	$\alpha_N = 0.5$
RFM	60.35	67.89	71.07	72.63	72.79	$\alpha = 1, \gamma = 2.5, \varepsilon = 0.8$

Table 4.1. Average retrieval rate (AVR) of 116 query images on DB1, obtained by automatic interactive learning. The initial AVR results (i.e., 0 Iter.) were obtained by Euclidean metric for ARBFN and single-RBF, and by Cosine measure for RFM.

The learning performances were also affected by the training samples, N_{RT}. With a fixed number of training samples, N_{RT} for every iteration, allowing many iterations, meant that performance deteriorated gracefully. For example, it was observed that the AVR results, obtained by ARBFN after 9 iterations, were reduced by 0.7% with FSD descriptor. This is because at later iterations most of the relevant images had already been found. Thus, if all relevant images are input into the SOTM algorithm as a training set, they will be split into two classes [according to Eq.(4.6)]; that is, misclassification will have occurred.

Fig. 4.8 provides an example of a retrieval session performed by the automatic ARBFN learning method, using FSD descriptor. Fig. 4.8(a) shows retrieval results without learning, and Fig. 4.8(b) shows the result after automatic learning. The improvement provided by the automatic retrieval method is apparent.

User Interaction Versus Semiautomatic Retrieval

In order to verify the performance of the automatic interaction learning of the MCI-CBR system, its performance was compared with that of HCI-CBR methods. The learning systems are allowed to interact with the user to perform the retrieval task, and the AVR results obtained are provided in Table 4.2. It was observed that user interaction gave better performance: 3.34% to 6.79% improvement after one iteration, and 3.66% to 4.74% after four iterations. However, it should be taken into account that the users had to provide feedback on each of the images returned by a query in order to obtain these results.

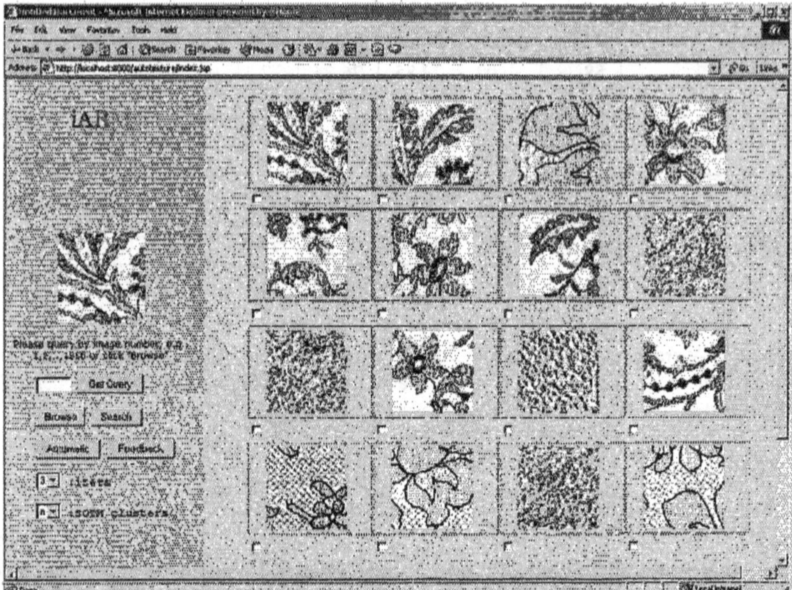

(a) Without interactive learning.

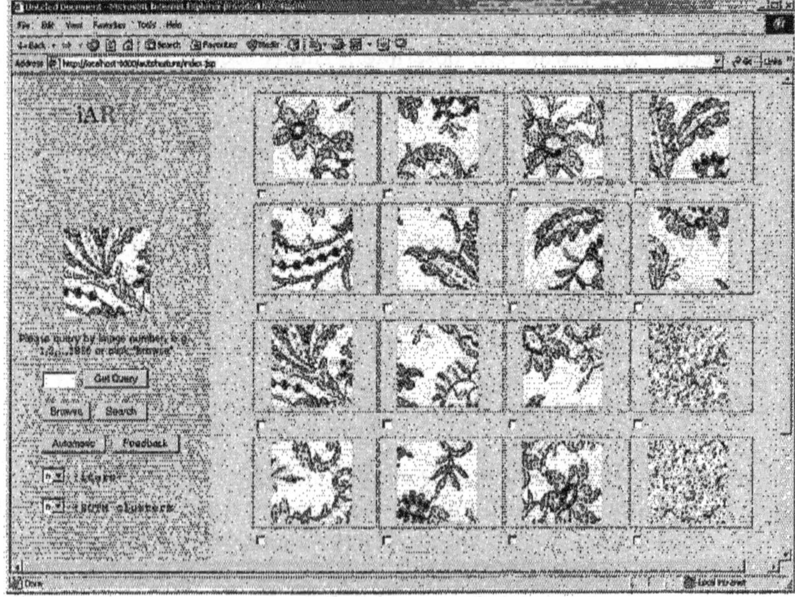

(b) After automatic interactive learning.

Figure 4.8. (a) Retrieval results without interactive learning; (a) retrieval results after application of automatic ARBFN.

As found in the studies from Chapters 3 and 4, retrieval performance can be progressively improved by repeated relevance feedback from the user. The semiautomatic approach reported here greatly reduced the number of iterations required for user interaction. This significantly improved the overall efficiency of the system. In this case, the retrieval system first performed an automatic retrieval for each query to adaptively improve its performance. After four iterations, the retrieval system was then assisted by the users. Table 4.3 provides the summary of the retrieval results, based on one round of user interaction. It was observed that the semiautomatic method is superior to the automatic method and the user interaction method. The best performance was given by semiautomatic ARBFN at 83.41% using FSD descriptors, and 81.14% using MHI descriptors.

In Fig. 4.9(a)–(c), the results is shown when each method reached convergence. The improvement resulting from the adoption of the semiautomatic approach is indicated by a correspondingly small amount of user feedback for convergence. In particular, the semiautomatic RFM, single-RBF, and ARBFN can reach or surpass the best performance of HCI-CBR within only one to two interactions of user feedback.

It is noticed that the superiority of the semiautomatic technique comes not only from the high initial result, but also from its use of the self-organizing method to identify other relevant samples. These are usually not easily presented by the user interaction method (i.e., it was possible to use a larger number of training samples, N_{RT} when working with the self-organizing method). This can be observed from the figures, where the semiautomatic methods converged with the higher AVR results, when these are compared to the user-controlled interactions.

User Subjectivity Tests

In a retrieval process, the term "similarity" usually suggests different things to different users. In this experiment, the automatic retrieval system was examined for its ability to deal with user subjectivity. Six users were invited to test the retrieval system. Each user was asked to judge the relevance of the 16 top-ranked images according to his/her own understanding and information needs. The judgments were made at two points: after the first round of retrieval, in order to evaluate the performance of the non-interactive CBR technique; and after the fourth round of retrieval, in order to evaluate the performances of

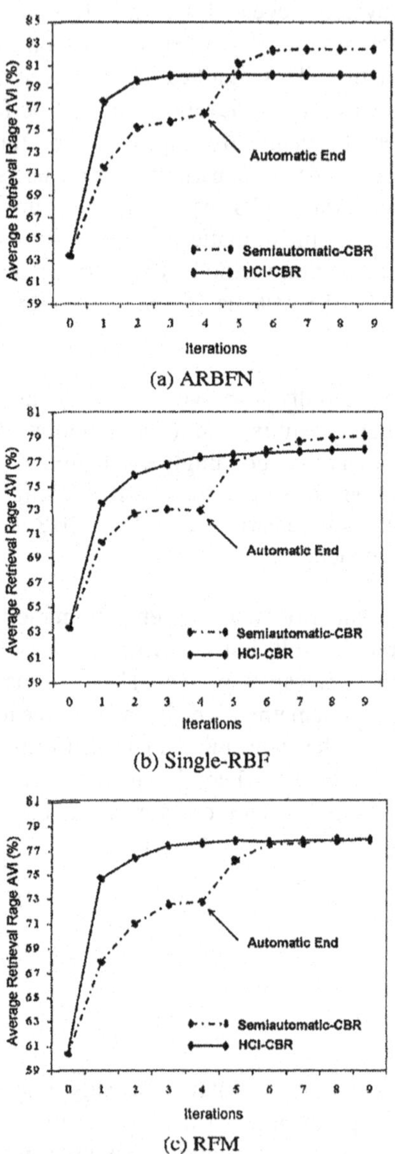

(a) ARBFN

(b) Single-RBF

(c) RFM

Figure 4.9. A comparison of retrieval performance at convergence, between the semiautomatic and HCI-CBR methods, where the similarity learning methods used are: (a) ARBFN; (b) Single-RBF; and (c) RFM. The semiautomatic method can attain the convergence within one to two iterations of *user feedback*. These results are based on the MHI descriptor.

Algorithm	Interaction Method	AVR (%)			No. of User RF (Iter.)
		0 Iter.	1 Iter.	4 Iter.	
ARBFN	a: MCI	63.42	71.66	76.51	-
	b: HCI	63.42	77.64	80.17	4
	$\Delta = b - a$	-	+5.98	+3.66	
Single-RBF	a: MCI	63.42	70.31	73.06	-
	b: HCI	63.42	73.65	77.43	4
	$\Delta = b - a$	-	+3.34	+4.37	
RFM	a: MCI	60.35	67.89	72.79	-
	b: HCI	60.35	74.68	77.53	4
	$\Delta = b - a$:	-	+6.79	+4.74	

Table 4.2. A comparison of AVR(%) between MCI-CBR method and HCI-CBR method, using DB1, and MHI descriptors, where Δ denotes AVR differences between the two methods.

Based on FSD descriptor			
Method	Initial Result	HCI-CBR	Semiautomatic CBR
ARBFN	58.78	77.53	83.41
Single-RBF	58.78	75.59	78.34
RFM	53.88	61.75	63.25
Based on MHI descriptor			
ARBFN	63.42	77.64	81.14
Single-RBF	63.42	73.65	77.05
RFM	60.35	74.68	76.39

Table 4.3. A comparison of AVR (%) between semiautomatic and HCI-CBR methods, using DB1, based on one round of user interaction.

automatic retrieval methods. All users were requested to provide similarity judgements, but not relevance feedback. No user had any *a priori* information about which retrieved images were in the same class as the given query. Thus relevance judgments were made only according to user subjectivity.

This system was implemented based on the ARBFN learning method for retrieval with the FSD descriptor. The test was conducted for all 116 query images, as used in the previous experiments. The AVR results examined by each user are summarized in Table 4.4. It was observed that the AVR average over all the users was 81.5% after four iterations, an improvement of 18% from the initial value of 63.5%. This result indicates that all users rated the automatic RF approach as the one which performed better than the one-shot

	Before Learning	After Automatic Learning
User1	66.60	83.14
User2	64.82	82.71
User3	64.39	82.60
User4	61.42	79.20
User5	63.41	83.46
User6	60.13	77.75
Average	*63.46*	*81.48*
Ground truth	58.78	77.21

Table 4.4. A comparison of AVR where the relevance judgment is based on ground truth and user subjectivity, based on ARBFN method, FSD descriptor, and DB1.

retrieval approach (in terms of capturing their perception subjectivity and information needs). *It also shows how the self-organizing machine understood the information requested based on a submitted query image.*

It was observed that AVRs fluctuated between 77.8% and 83.5% according to the users. The level of fluctuation is even more dramatic when we compared the AVRs from each user to the one based on the ground truth criterion. This was to be expected, since no fixed similarity measure can cope with different relevant sets across users.

Retrieval in the DCT Compressed Domain

In this section, the ARBFN learning method is applied, in both automatic and semiautomatic modes, for retrieval of images in a JPEG photograph database [38]. The energy histograms of the lower frequency DCT coefficients [39] (described in Section 3) are used to characterize each image in the database. The four coefficients bounded within the upper left corner of the DCT block are used to obtain image descriptors. Separate energy histograms are constructed for the DC and AC coefficients of each of the color channels, and 30 bins are used for each histogram. For relevance classification, the GWT [24, 88] is again adopted to characterize the retrieved images. The GWT was applied to the dominant colors in each channel and the transform coefficients were used to construct a descriptor. This gives better characterization of texture information from different color spaces. In each interactive session, the SOTM was used to identify image relevance on the 18 top retrievals. The performance was measured on the top twelve retrieved images after four iterations.

Table 4.5 provides average *relative precision*[4] (APR) figures for thirty query images (show in Fig. 4.7). In general, conclusions similar to those for the texture database can be drawn from these results, with regard to the retrieval performance. The semiautomatic method consistently displayed superior performance over the other methods discussed: improvement was from 49.8% to 98.1%, with the number of user feedbacks reduced by half to reach convergence.

The retrieval session for this database is shown in Fig. 4.10(a)-(b). Fig. 4.10(a) shows the 12 best-matched images without learning, with the query image displayed in the top-left corner. It is observed that some retrieved images are similar to the query image in terms of texture features. Note that seven similar images are relevant. Based on this initial information, the self-organizing system dynamically readjusts the weight parameters of the ARBFN model to capture the notion of image similarity. Fig. 4.10(b) displays the retrieval results, which are considerably improved after using the automatic interactive approach. Fig. 4.11(a) shows the retrieval results of the semiautomatic ARBFN in comparison to the user-controlled interaction illustrated in Fig. 4.11(b). It can be observed that semiautomatic results are visually better than those of the direct user-controlled method. The semiautomatic method reduced the number of user interactions by half to reach convergence.

In Fig. 4.12(a), the query image in the top-left corner depicts a sunset scene. In the first iteration, there are six retrieved images which are similar to the query. These images are then incorporated into the training set. After automatic retrieval (Fig. 4.12(b)), more images depicting sunsets and related images, showing a red sky background, are returned. Fig. 4.13(a) shows the results from the semiautomatic learning mode, which can be compared to the results of the user-controlled interactive method shown in Fig. 4.13(b). It can be seen that the semiautomatic method shows improvements in retrieval accuracy as well as convergence speed.

6. Chapter Summary

In this chapter, a novel framework is presented for automatically retrieving images in digital video libraries and applied to image retrieval in compressed domains. It was shown that, by introducing a second feature space of very high quality, the SOTM can be used for selecting relevance in a flexible fashion. This minimizes user participation in an effort to automate interactive retrieval.

(a) Non-interactive CBR, $Pr = 0.58$.

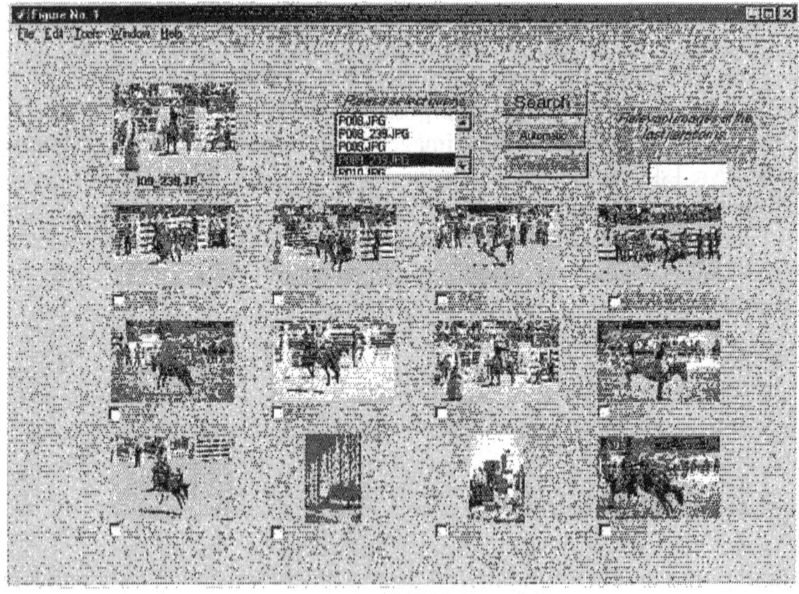

(b) Automatic ARBFN, $Pr = 0.83$.

Figure 4.10. Retrieval results from DB2, obtained by (a) noninteractive-CBR; (b) automatic ARBFN.

(a) Semiautomatic ARBFN, $Pr = 1$.

(b) User-controlled ARBFN, $Pr = 1$.

Figure 4.11. Retrieval result from DB2, obtained by (a) semiautomatic ARBFN, converted by one round of user interaction, (b) user-controlled ARBFN, converted by two rounds of user interaction.

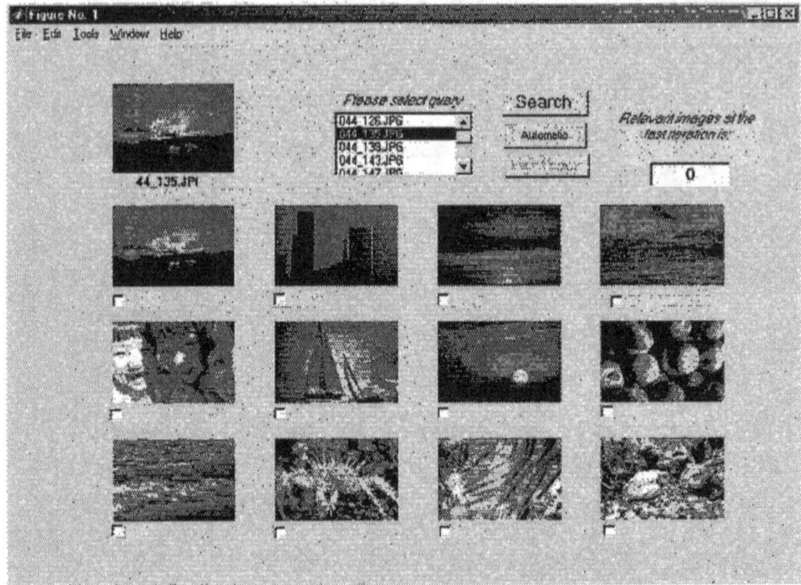

(a) Non-interactive CBR $Pr = 0.5$.

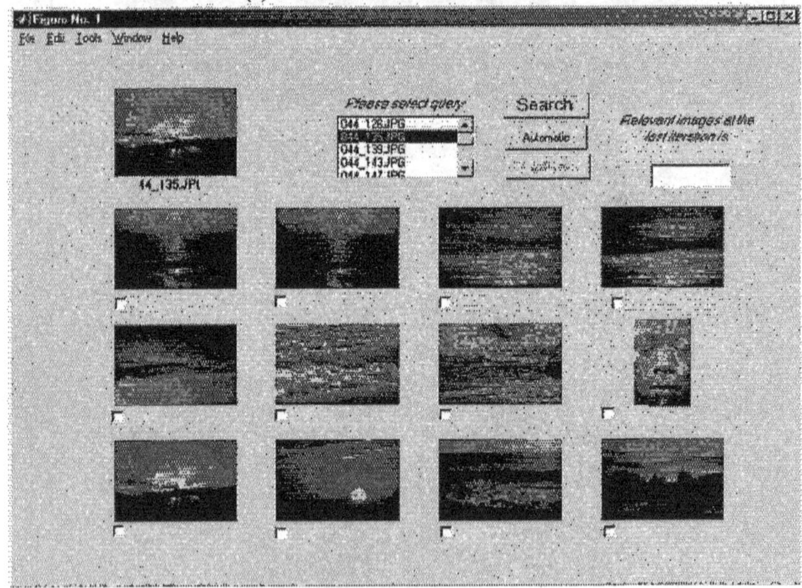

(b) Automatic ARBFN $Pr = 0.92$

Figure 4.12. Retrieval results from DB2, obtained by (a) noninteractive CBR; (b) automatic relevance feedback.

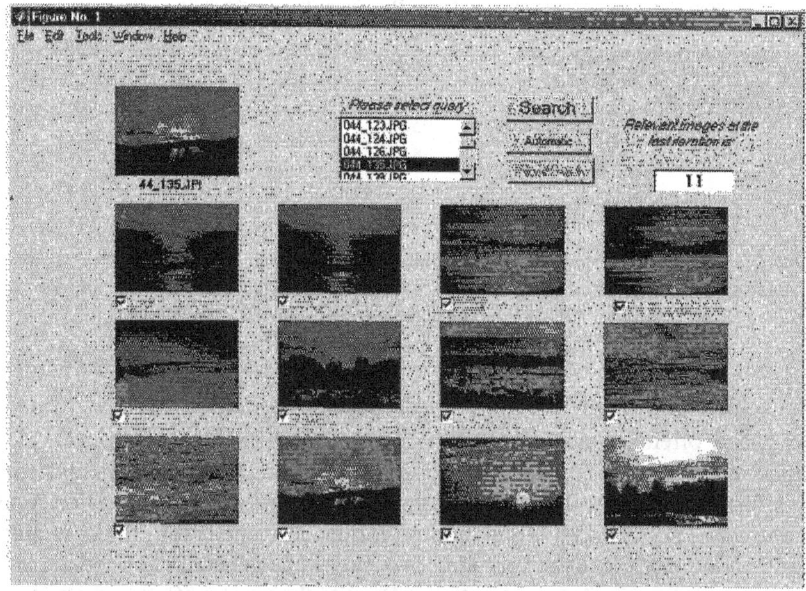

(a) Semiautomatic ARBFN, $Pr = 1$.

(b) User-controlled ARBFN, $Pr = 1$.

Figure 4.13. Retrieval results from DB2 database, obtained by (a) semiautomatic ARBFN, converted by two rounds of user interaction (b) user-controlled ARBFN, converted by three rounds of user interaction.

Method	Avg. Relative Precision (%)	Avg. no. of User RF for Convergence (Iter.)
Non-interactive CBIR	49.82	-
MCI-CBR	79.18	-
HCI-CBR	95.66	2.63
Semi-Automatic CBR	98.08	1.33

Table 4.5. Retrieval results on DB2: Column 2: Relative precision (%); Column 3: Number of user feedbacks (iterations) required for convergence, averaged over 30 query images. The results were performed by ARBFN in both automatic and semi-automatic modes.

The self-organizing architecture indeed provides a unified automatic framework, and offers an attractive facility for practical interactive retrieval. The structure can also be embedded naturally in semi-automatic retrieval systems. This further enhances the ability of the system to model user needs on a per query basis. Based on a simulation performance comparison, the new automatic system appears to be very effective and can be generally applied to many compressed-domain feature structures. Despite the proven effectiveness of automatic relevance feedback strategies, which can handle large sample numbers, semiautomatic retrieval systems are still slightly more accurate in the ARBFN and single-RF modes until convergence occurs.

Notes

1 Although not a standard, SPIHT is a well-known representative of state-of-the-art wavelet codecs and serves as a reference for comparison [112].

2 The Brodatz database has been previously used in the experiments in Chapter 2. It contains 1856 texture images classified into 116 classes.

3 Note that a feature normalization algorithm was applied to each feature database.

4 Relative precision is defined as $\frac{Pr}{\max Pr}$. In this experiment, Pr is the number of relevant images over twelve. This provides an easy way to compare the relative performance among different retrieval methods.

Chapter 5

AUTOMATIC RETRIEVAL ON P2P NETWORK AND KNOWLEDGE-BASED DATABASES

1. Introduction

The previous chapter described how self-organizing network is used to automate relevance feedback process. The SOTM model is independent of the image description(s) used to implement image retrieval; it focuses attention on the requirements for successful relevant classification. The automatic relevance feedback was conducted on compressed image databases. But this still begs the questions of how the automatic relevance feedback architecture can be adjusted to various domain applications and how to produce new ways of accessing digital libraries. This chapter addresses this issue further by firstly conducting automatic relevance feedback for distributed databases using a peer-to-peer communication structure (Sections 2-3). Then, it explores the automatic retrieval framework by incorporating the use of knowledge to produce some levels of the equivalent classification performance used in human vision (Section 4).

2. Image Retrieval in Distributed Digital Libraries

In the internet era, rich multimedia information is easily accessible by everyone attached to the network. While the amount of multimedia content keeps growing, locating the relevant information becomes more and more time consuming. A successful networked CBR system should therefore maximize the retrieval precision while optimizing the network resources. Most of the CBR systems proposed today assume a centralized query server, such as QBIC [1] and VisualSEEk [2]. Alternatively, content query over distributed peer-to-peer (P2P) network was studied in [145]. This assumes that a peer contains only

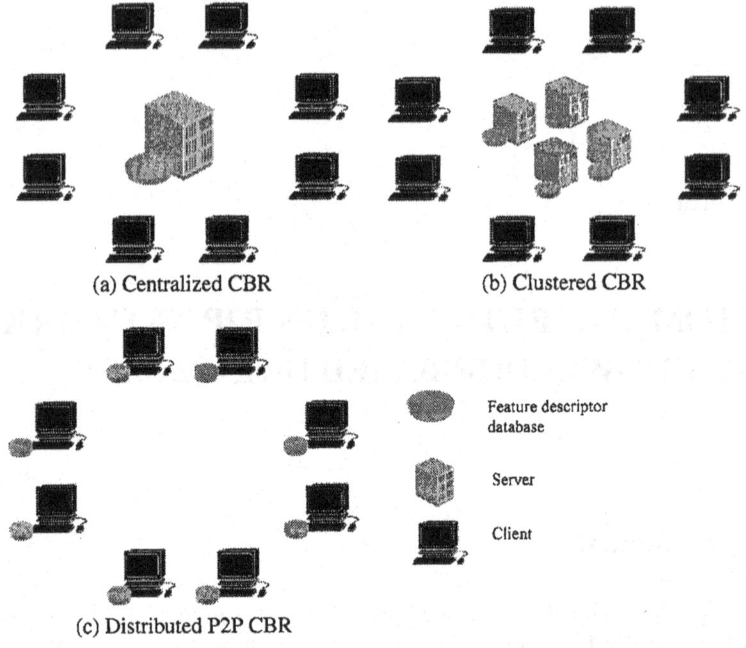

Figure 5.1. Networked CBR system.

one image category. In [146], a meta-search engine (MSE) incorporating the search-agent processing model (SAPM) was studied to handle the search across a substantial number of distributed databases. In this chapter, the work initiates a different approach by studying practical scenarios where multiple image categories exist in each individual database in the distributed storage network. In this scenario, an automatic relevance feedback method described in Chapter 4 is applied to the networked databases, where the system can achieve high retrieve accuracy as well as minimize network resources.

CBR over Centralized, Clustered, and Distributed Storage Networks

According to the distribution of the feature database, CBR systems can be classified into centralized, clustered, and distributed P2P systems, as illustrated in Figure 5.1.

Centralized CBR System

A centralized CBR system maintains a central server to handle the query requests. Upon retrieving the relevant images according to the feature similarity measure, the universal content locator (URL) will be returned to the requesting host. The actual content will be transferred directly from the content server to the requesting host. The centralized CBR systems keep the entire feature descriptor database in a centralized server. The real image content may or may not be located on the same server. The centralized CBR server retrieves relevant content based on the feature-descriptor database. The drawback of the centralized CBR system is the scalability to handle growing retrieval requests and larger image databases.

Clustered CBR System

When the image database grows, the centralized CBR approach does not scale to handle the growing computational requirement. Therefore, a clustered CBR system is studied here to pre-classify the feature descriptor database into non-overlapping categories. Each category is stored in a different server. The clustered CBR approach helps to reduce the computational and bandwidth requirement for the centralized CBR system. To address the issue of scalability for centralized CBR system, the feature descriptor database can be pre-clustered and stored in different servers. Each server in the cluster will pre-compute the centroid, which is the mean of the feature descriptors stored in this particular server.

During the query stage, the best query server is firstly identified from the similarity measure of the query feature descriptor and the cluster centroid using the nearest neighbor algorithm. Once the query server is identified, relevant content will be retrieved within the feature descriptor database stored on this server.

In this study, only the best server is identified to perform the content query. Content query over multiple best-matched servers may potentially result in better retrieval precision at the cost of computation and bandwidth.

3. Peer-to-Peer (P2P) Retrieval System

To further acknowledge the high correlation between each individual image database resulting from, for example, hobbyist photo collections, a decentralized CBR system attempts to group distributed nodes which share the same image categories. Then, retrievals are made at each node. A special case of the decentralized database system using the peer-to-peer (P2P) network is studied. Each node in the P2P network acts both as a client for requesting images and a server for re-distributing the images. Since a peer can join and leave the network at any time, a challenge of this distributed CBR system is to address the non-guaranteed level of service of the P2P network. To localize the search, the query packet is always associated with certain Time-to-live (TTL) levels. Database storage on distributed servers had been applied in the industry to provide high availability (providing continuous service if one or more servers are unintentionally out of service) and efficiency (access from the closest server geographically). P2P network is a special case of such a network where each node in the network behaves as a database server.

The underlying assumption of the multimedia content collection on a P2P node consists of limited categories, as the user behavior determines the data collection. Distributed CBR systems can benefit from such high correlation among certain peers, hence reduce the computation and bandwidth cost by searching within a limited subset of peers. Each peer's image collection can be considered as a subset of the whole image database, and no assumption is made on the inter-dependencies or collaborating work between any two peers. Therefore, the overlap between peers can be used to improve the retrieval precision.

Each peer in the P2P CBR system maintains two tables of neighbors. The first type of neighbors are called the *generic neighbor* which typically represent the neighbors with the least physical hop counts. The other type of neighbors are called the *community neighbors* and common interest is shared among the community. Two stages of operation are required: *community neighborhood discovery* and *query within the community neighborhood.*

Community Neighborhood Discovery

As shown in Figure 5.2(a), a peer node originates the query request to its generic neighbors in the P2P network. Whenever a peer node receives a query request, it will (1) decrement the TTL, and forward the request to the generic neighbors when TTL $>$ 1, and (2) perform the content search within the peer's feature descriptor database. The retrieval results of each peer are transmitted to the original query peer directly in order to improve the efficiency.

Like most P2P applications, the proposed distributed CBR system applies an application layer protocol, such that the system can be realized on today's internet without modifying the underlying network infrastructure. The proposed query packet format to traverse through the P2P network is shown in Figure 5.2(c).

Once the destination peer receives the query and performs feature match, it will issue a Query Reply to the query requester directly. The query results are in the form of filenames and distance. The actual file transfer is not part of the protocol, and protocols like HTTP, RTP, with or without encryption, may be applied depending on the application. Transferring the actual image content is coupled with the feature descriptor transmission, to eliminate the need to re-compute the feature descriptors upon receiving a new image. The query search and query response packet format to traverse through the P2P network are shown in Figures 5.2(b) and 5.2(d), respectively.

The query peer maintains a table of community neighbors based on past retrieval results to identify the peers which collects similar image database.

Query within the Community Neighborhood

Once the community neighbors are identified, subsequent queries will be made to limited peers within the community neighborhood. To improve the communication efficiency, instead of forwarding the request hop-by-hop in the community neighborhood discovery stage, direct communication between the peers is applied. The same packet format is used for Query and Query Response within the community neighborhood.

Each peer in the community neighborhood collects more than one category of images, with at least one common category as the requesting peer to satisfying the criteria to be listed in the community neighborhood. Therefore, the same image appears in multiple peers are likely belong to the common category in the community neighborhood. Let $Ret(I,P_n)$ denote the retrieved result using query image I from peer P_n, where $P_n \in \{\text{community neighborhood}\}$. Let $N(\cap Ret(I,Pn))$ denote the number of occurrences of each retrieved image I. Let $D_{N(I)}$ be the occurrences distance, which is calculated by normalizing $N(\cap Ret(I,Pn))$. Assign the weighting factor $Wp2p = [W_D \; W_N]$ to the feature distance $D_{feature}$ and distance measure according to the number of occurrences $D_{N(I)}$, respectively. The similarity ranking is:

$$Rank = W_{p2p} \cdot [D_{feature} \; D_{N(I)}]^T \tag{5.1}$$

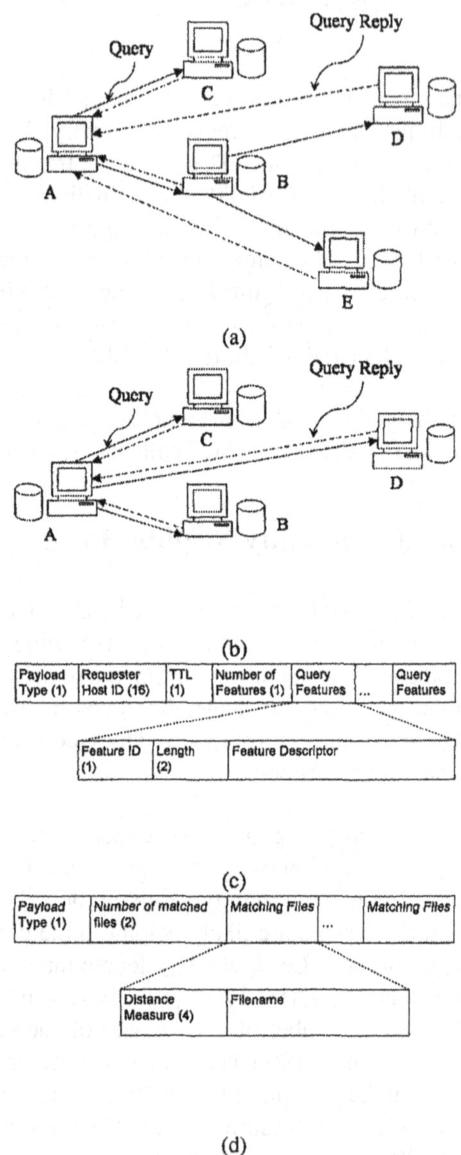

Figure 5.2. (a) Neighborhood discovery, (b) Search within community neighborhood, (c) Packet format for Query, (d) Packet format for Query Reply

Distributed CBR with Automatic Relevance Feedback

While automatic relevance feedback (ARF) reduces the need for human interaction for relevance feedback, integrating ARF to the distributed CBR framework introduces new challenges for repeated requests to multiple peers, which consumes bandwidth and computational resources. To address this issue, an incremental searching mechanism is conducted to reduce the level of transactions between the peers.

As shown in Figure 5.3(a), peer A originates a query request to its nearest neighbor peer B. Peer B performs the query, and returns the top matched feature descriptors to peer A. Consequently, Peer A evaluates the retrieval results using the self-organizing tree map (SOTM) algorithm, and generates a new feature vector using RBF method. A new query request using the new feature vector will be sent to peer B, as well as incrementing the audience to peer C. The query request and automated retrieval evaluation process is repeated until a pre-defined number of query peers is reached.

Offline Feature Calculation

As described previously, online feature calculation requires high computational resources and results in delay for content retrieval. Redundant online feature computation can be eliminated by the following specifications:

- Each image stored in the P2P CBR network is attached with its feature descriptor.
- When a peer creates a new image, the feature descriptors will be computed and attached with the image file before announcing the availability of the new image.
- Any image transmission over the distributed P2P network will be coupled with the transmission of the image's feature descriptor.

Advanced Feature Calculation

For extensibility, apart from offline feature calculation, an ideal CBR system should also allow new features to be computed on the fly. These new features, typically address the specific feature for a query image, are best implemented with the ARF such that no additional human interaction is required. To realize advanced feature descriptors for the distributed CBR systems, two approaches are studied: Query Node ARF and Agent-based ARF:

Query Node ARF: As illustrated in Figure 5.3(a), the query node, Peer A, makes the initial request to the destination node, Peer B. The destination nodes will transfer the retrieved images to the query node. The query node will calculate the advanced feature descriptor on the fly and perform ARF. A new query vector will be generated and a new query will be made from the query node to the destination nodes repeatedly, until a pre-defined number of iterations is met.

Agent-based ARF: The major drawback of the Query Node ARF approach is the bandwidth cost for multiple retrieved image transmission to the query node, as well as the computation cost for the query node to calculate advanced features on-the-fly. We propose an infrastructure applying the software agent technique [147] to offload the bandwidth and computation cost from the query node. As shown in Figure 5.3(b), the query node, Peer A, initiates a software agent to carry the query vector using a standard feature descriptor, as well as the advanced feature algorithm, to the destination node, Peer B. Peer B performs the retrieval with ARF with the query vector together with the advanced features computed for all the images on-the-fly. The software agent carries the query vector, advanced feature algorithm, as well as the retrieved images from Peer B, to the subsequent neighbor node, Peer C. Upon reaching a pre-defined number of neighbors, the software agent will carry the retrieved images back to the query node.

Offloading computational cost from the query node to destination nodes raises security concern, as the flexibility of remote procedure execution opens the doorway for various malicious attacks. Therefor, authenticating as well as validating the integrity of the software agent is a must.

Experimental Results

For experimental purposes, a P2P network is constructed using an evenly distributed tree structure and each peer is connected to five other peers. The number of image categories each peer possesses follows normal distribution, with the mean μ_{cat}=10 and the standard deviation σ_{cat}=2. The number of image per category is also normally distributed, with the mean μ_{images} =50, and the standard deviation σ_{image}=5.

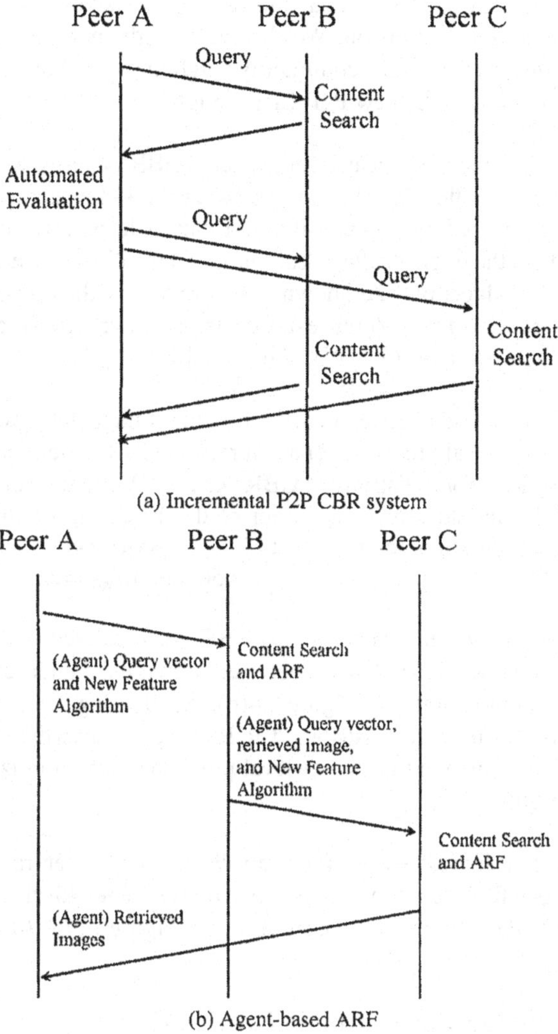

(a) Incremental P2P CBR system

(b) Agent-based ARF

Figure 5.3. Advance Feature Calculation: (a) Incremental P2P CBR system (b) Agent-based ARF.

The simulation is performed with the Corel photo image database, which consists of 40,000 color images. The statistical results are taken from averaging the 100 queries in the categories of bird, canyon, dog, old-style airplane, airplane models, fighter jet, tennis, Boeing airplane, bonsai, and balloon.

Figure 5.4(a) shows the statistical analysis of the size of community neighborhood to the retrieval precision. We observe a steady increase for the retrieval precision against the size of the community neighborhood. Such characteristics serve the foundation of the new P2P CBR system.

The same experiment is repeated for clustered CBR systems, where the image database is pre-classified into 10 clusters using the k-means algorithm. Prior to the similarity matching process, the best cluster to perform the retrieval is determined from the query feature descriptor and the cluster centroid using the nearest-neighbor algorithm. As shown in Figure 5.4(b), the proposed clustered CBR system trades off 5.75% retrieval precision in average for offloading the computation with an order $O(1/N)$ where N=10.

The inter-dependence between each individual image database can be used to improve the retrieval precision for a centralized CBR system, using the same algorithm proposed for a distributed CBR system. While the centralized CBR system typically includes a higher order of database, higher diversity is expected. In this simulation, the number of images per category is also normally distributed, with μ_{cat}=20, σ_{cat}=5, μ_{images} =50 and σ_{image}=20.

Comparisons between the centralized CBR, the clustered CBR, centralized CBR accounting inter-dependencies between individual databases, and the distributed CBR, are illustrated in Figure 5.4(b). Accounting the overlap between relevant databases used for distributed P2P CBR, as described in Section 3, we observed improvement in the retrieval precision for centralized CBR with similarity weighting.

Finally, screen shots of a query for an airplane, from the centralized database, with the first and fifth iteration of interactive relevance feedback, and with ARF on the distributed P2P CBR system, are shown in Figure 5.5(a)-(d), respectively.

4. Knowledge-Based Image Retrieval

It is well acknowledged that to obtain maximum precision rate in image retrieval, it is critical that the RF learning process should effectively exploit the knowledge of image relevancy. This requires modeling image contents with sufficiently accurate features for the characterization of perceptual importance. As discussed in the previous section, the general principle of automatic RF is to apply unsupervised learning techniques to relevance classification in order to minimize the number of user feedback cycles required in modeling user's

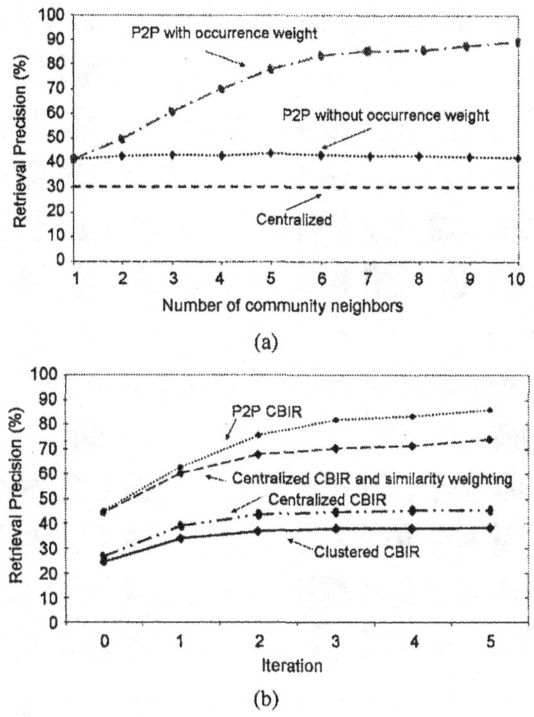

Figure 5.4. (a) Statistical retrieval result for the proposed P2P CBR system, (b) CBR with automatic relevance feedback using RBF and SOTM methods.

queries. However, the importance of the perceptually inspired features in the relevance classification process has not been properly studied.

This issue is especially pressing with automatic relevance feedback since, without providing some form of knowledge to the relevance classification process from the external world, the SOTM classifier cannot operate as efficiently as that of a user supervision process. For example, global features of shape, color, or texture information might consume an undue proportion of weights toward the judgment of image relevancy by machine vision. Furthermore, these global features do not always address perceptually important regions or any salient objects depicted in an image. This is because there are more regions in an image than those which are of perceptually importance. So, higher classification accuracy may be possible with the acquisition of more precise perception information. However, the form of knowledge needed in automatic relevance

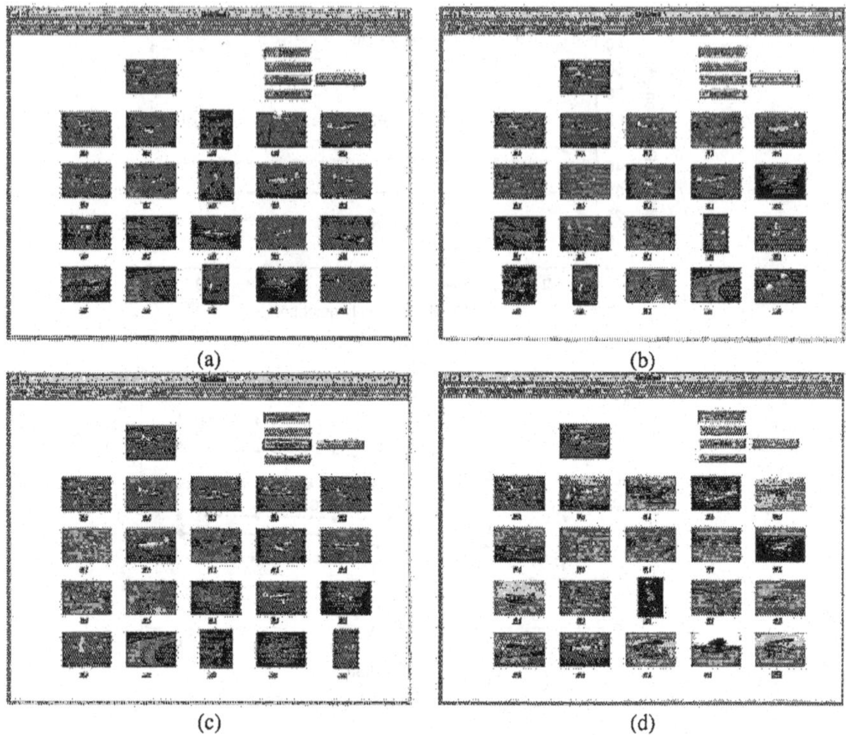

Figure 5.5. (a) CBR from a centralized database, the retrieval precision is 55%, (b) Relevance feedback with single iteration, the overall retrieval precision is 70%, (c) Relevance feedback with five iterations, the overall retrieval precision is 80%, (d) P2P CBR with automated relevance feedback, the overall retrieval precision is 95%

feedback has to be identified before the retrieval process begins, instead of during the process as the user-interactive relevance feedback does.

In this section, an automatic adaptive image retrieval scheme is implemented with embedded knowledge of perceptual importance, the form of which is identified in advance. With a specific domain to photograph collection, we pursue the restricted goal of identifying the region of interest (ROI). The ROI assumes that the significant objects within an image are often located at the center, as a photographer usually tries to locate significant objects at the focus of the camera's view. The Edge Flow model [148] is adopted to identify the ROI within a photograph. This ROI does not necessarily require the exact identification of a possible object in the image, but only the region selected which adequately reflects those properties of the object such as color or shape which are usually used as features for matching in retrieval.

Characterization of the ROI

Image segmentation is considered as a crucial step in performing high-level computer vision tasks such as object recognition and scene interpretation [149]. Since natural scenes within an image could be too complex to be characterized by a single image attribute, it is more appropriate to consider a segmentation method that is able to address the representation and integration of different attributes such as color, texture, and shape. The Edge Flow model demonstrated in [148] is adopted, which proved to be effective in image boundary detection and application to video coding [150]. The Edge Flow model implements a predictive coding scheme to identify the direction of change in color, texture, and filtered phase discontinuities.

Region characterization by "Edge Flow"

An edge flow vector at pixel location s is a vector sum of edge energies given by:

$$\vec{F} = \sum_{\Theta(s) \leq \theta \leq \Theta(s)+\pi} E(s,\theta) \cdot \exp(j\theta) \tag{5.2}$$

which is taken along a continuous range of flow directions that maximizes the sum of probabilities:

$$\Theta(s) = \arg\max_{\theta} \left\{ \sum_{\theta \leq \theta' \leq \theta+\pi} \mathrm{P}(s,\theta') \right\} \tag{5.3}$$

where $\mathrm{P}(s,\theta)$ represents the probability of finding the image boundary if the corresponding Edge Flow "flows" in the direction θ. The model in Eq. 5.2 facilitates the integration of multiple attributes in that each Edge Flow obtained from different types of image attributes:

$$E(s,\theta) = \sum_{a \in A} E_a(s,\theta), \tag{5.4}$$

$$\mathrm{P}(s,\theta) = \sum_{a \in A} \mathrm{P}_a(s,\theta), \tag{5.5}$$

where $a \in$ {intensity/color, texture, phase}.

The basic idea of this model is to identify the direction of change in the attribute discontinuities at each image location. The Edge Flow vectors propagate from pixel to pixel along the directions being predicted. Once the propagation process reaches its stable state, the image boundaries can be detected by identifying the locations which have non-zero edge flows pointing to each other. Finally, boundary connection and region merging operations are applied to create close loop regions and to merge these into a small number of regions according to their color and texture characteristics.

Region of Interest (ROI)

The definition of ROI is highly dependent on user needs and perception. For example, in a medicine application, a doctor may find various objects as ROI in medical images such as regions with clustered micro-calcifications in a digital mammogram. However, specific to the current application for photographic collections, a photographer usually creates a photograph with a single focus point at the center of the picture. Based on this post production knowledge, we can effectively attain ROI by associating it with the objects that locate at the center of photographs. Let $S = \{R_i,\ i = 1, 2, ..., N | R_a \cap R_b = \phi\}$ be a set of regions generated by the Edge Flow model from one image. Also, let W be a set of labels of regions that locate either partly or completely inside a predefined m-by-n pixel window, where the center of the window is located at the center of the image. ROI is defined as a collection of regions which are the members of W:

$$S' = \bigcup_{i \in W} R_i \tag{5.6}$$

In practice, however, it is important to remove from S' the background or other regions that are located mainly outside the defined m-by-n window. In this case, a region R_i, $i \in W$, is not included in S' if its area outside the window is greater than its area inside the window. Thus, a complete region of interest is defined as:

$$ROI := S' \bigcap_{i^*} comp(R_{i^*}) \tag{5.7}$$

where *comp* is the complement set operation. In this definition, each R_{i^*} is defined so that:

$$R_{i^*} = R_i \text{ if } M < N \tag{5.8}$$

where M and N are the number of pixels of R_i that locate inside and outside the m-by-n window, respectively.

Figure 5.6 shows some image examples after applying ROI characterization process to four images from Corel digital collection [22].

5. Relevance Feedback with Region of Interest

Figure 2 shows the diagram of automatic relevance feedback system with embed knowledge of ROI. Automatic relevance feedback is performed in two stages. First, the SOTM is applied for relevance classification to label the retrieved samples as positive or negative. Then, the labeled samples form the input to a non-linear similarity ranking function based on a single-class RBF network.

Let $\{X_i\}_{i=1}^{K}$, $X_i \in \Re^{p_1}$ be the set of samples retrieved at the first round of retrieval, subjected to:

$$D_1^{(Z)} \leq D_2^{(Z)} \leq \cdots \leq D_N^{(Z)} \tag{5.9}$$

where $D_i^{(Z)}$ denotes the distance between sampleX_i and the query Z. Here, the query Z and the samplesX_i are each represented by a point in the feature space $\Re^{p_1}$ and consist ofp_1 features, including color, shape, and texture. As discussed in Chapter 4 that in order for the unsupervised learning process brought into automatic relevance feedback to be effective, a different and more powerful feature space other than $\Re^{p_1}$should be introduced in relevance classification. Here, the feature space $\Re^{p2}$ is used. Apparently features in $\Re^{p2}$ should be of higher quality than those in $\Re^{p_1}$ in order for the relevance identification process to be effective. Because of the limited number of images being analyzed (only those identified by the retrieval process as "similar" to the query), it is affordable to employ computationally more intensive but also more effective image processing/analysis methods in order to obtain features which are perceptually important in relevance classification. Apparently features extracted from the ROIs satisfy this requirement as they provide the embedded knowledge of ROI in assisting relevance classification. Therefore, a different set of features, $\left\{\tilde{X}_i\right\}_{i=1}^{K}$, $\tilde{X}_i \in \Re^{p_2}$ is extracted to characterize the retrieved images. Color and shape are again chosen as the features, but they are calculated only from the ROIs in the retrieved images after applying ROI identification.

Retrieval Results

Results provided in this section were obtained using the Corel Digital Library [22], together with its pre-defined ground truth classes. Although some

Figure 5.6. Characterization of Region of Interest (ROI) using Edge Flow model and centering window method. *Left column*: Edge Flow boundary detection; *Middle column*: ROI identification; and *Right column*: ROI image results.

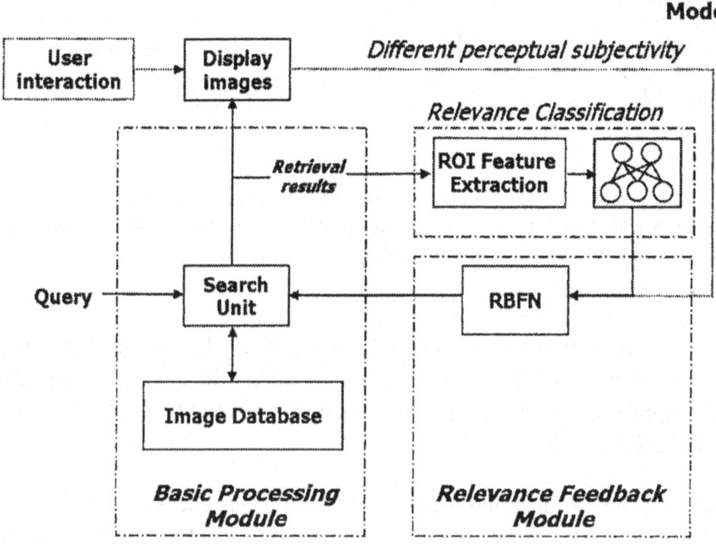

Figure 5.7. Diagram of an automatic relevant feedback image retrieval system.

classes of images in this database are overlapping based on the ground truth created by Corel Professionals, further judgment was conducted by a user to provide fair results. The selected database has more than 11,500 JPEG files, covering a wide range of real-life photos.

Each image is indexed by 48-bin HSV color histogram and color moments for color descriptors; Gabor wavelet method for texture; and Fourier descriptor for shape [151]. This produces a 115-dimensional feature vector X_i which is used for retrieval. For the identification of image relevancy by SOTM, the retrieved images are passed through Edge Flow algorithm to identify ROIs, followed by extracting the feature vector $\tilde{X}_i$ within the ROIs, using color histogram and Fourier descriptor. Here, we also pay attention to the application of a weighting scheme to the color and shape descriptors which are the input to the SOTM, to embed user's perception in the SOTM relevance identification process.

The performance comparisons were conducted using four methods: non-adaptive CBIR, user-controlled RF, automatic RF, and semi-automatic RF, using 20 queries from different categories. The non-adaptive CBIR method employs normalized Euclidean distance as matching criterion. This method provides a retrieved image set to the user-controlled RF algorithm that further enhances the system performance by the non-linear RBF model, together with user interaction

Method	Average Precision (%)	# User Feedback
Non-adaptive CBIR	52.81	-
Automatic RF	78.13	-
User-controlled RF	82.19	1
Semi-automatic RF	87.50	1

Table 5.1. Average precision rate (%) and number of user feedback cycles, obtained by retrieving 20 queries from Corel database, measured from the top 16 best matches. Precision is defined as $n_p/16$, where n_p is the number of relevant images retrieved.

interface. In comparison, in automatic RF case, the relevance identification was executed by the SOTM with two cycles of adaptation. In addition, after automatic adaptation, the system performance was refined by a user to obtain semi-automatic RF results.

Table 5.1 presents results obtained by the four methods, measured by the average precisions on the top 16 best matches. Evidently, the automatic RF provides considerable improvement over the non-adaptive CBIR method (i.e., by more than 25% in precision measure), without user interaction. The automatic result is ≈4% lower than that of user-controlled RF method. By combining automatic learning with user interaction, it is observed that the semi-automatic RF clearly outperforms others methods discussed.

The user interaction process was also allowed to continue until convergence. It is observed that the user-controlled RF and the semi-automatic RF reached the convergence at the similar points at ≈ 93% in precision measure. However, in order to reach this optimum point, the user-controlled RF method used on average 2.4 cycles of user interaction, while the semi-automatic RF method used 1.6 cycles. This shows that the semi-automatic method is the most effective learning strategy in both retrieval accuracy and minimizing user interaction. This also demonstrates that the application of self-organizing RF in combination with perceptually significant features extracted from the ROIs clearly enhanced the overall system performance.

Discussion

In the context of a relevance feedback retrieval framework, the user workload and processing time introduced by user-interaction are relatively high due to a series of querying processes. With the proposed automatic relevant feedback strategy with perceptually important features extracted from the regions of interest, user interaction can be eliminated from relevant feedback image retrieval. Simulation results demonstrate that the unsupervised relevance feedback learning strategy based on the self-organizing tree map effectively utilized the perceptually important features extracted from the ROIs in relevance classification and substantially improved retrieval accuracy without user interaction. In addition, it is observed that it is possible to find the optimal tradeoff between the automatic process and the number of feedback cycles required by the user in order to achieve very high retrieval accuracy.

Chapter 6

ADAPTIVE VIDEO INDEXING AND RETRIEVAL

1. Introduction

Accurate characterization of visual information is the primary requirement for constructing an effective content-based video retrieval (CBVR) system. Consequently, it is important to develop analysis techniques specific to the time-varying nature of video, rather than relying upon the application of still-image techniques. The last part of this book presents video indexing and retrieval techniques specific to video databases. This chapter firstly demonstrates an adaptive video indexing (AVI) technique that allows full use of video content analysis in temporal dimension. Then, it will discuss the integration of this representation within a neural network structure able to adaptively capture different degrees of visual importance in a video sequence, increasing retrieval accuracy.

In most systems [40, 41], video retrieval is done by first breaking up the video sequences into temporally homogeneous segments called shots. These segments are then condensed into one or a few representative frames which are then used to determine the similarity between shots on the basis of their visual characteristics. While this approach constitutes an efficient and valid CBVR system, it has proven to be inadequate in many ways. Among these is the extensive use of dynamic visual content. In most cases, it is difficult to extract information from single frames since their meaning is sequential. This results in their becoming meaningless when taken out of context. Furthermore, it is not always the case that events or actions can be accurately attributed based on a single frame [42].

This chapter presents an adaptive retrieval system that places a strong emphasis on embedding representations of the dynamic contents in video. An adaptive video indexing (AVI) technique is developed to encode all visual contents occurring in the video. Video data is viewed as a collection of visual templates and so capture sequential information by a specific weight vector. In contrast to key-frame based video indexing (KFVI) techniques, AVI takes into account the importance of all frames in the video sequence through a differential weighting algorithm. By doing so, AVI is highly adaptable to support indexing beyond the shot level, through the use of finely embedded representations of the video. This provides flexibility in accessing video collections from a multi-level perspective. In other words, the AVI system offers users the option to retrieve a video scene or story, using video clips that contain more accurate narratives.

The AVI takes into account video indexing along temporal dimension. Thus, it increases effectiveness of video content analysis via relevance feedback algorithm. The implementation of this can be achieved through automatic relevance feedback (discussed in Section 5). In an audio-visual fusion processing system, AVI is employed to characterize visual information combined with audio content analysis module. This fusion model increases capability of the system in retrieving video data when user submits a concept-based query discussed in Chapter 7.

Video Indexing and Retrieval

The previous studies on video indexing and retrieval include video segmentation (e.g, shot, scene, story segmentation), video-content representation, and automatic detection and recognition of objects and events. First, a segmentation algorithm is applied to the video data to obtain video intervals. Then, content characterization techniques are applied to the video intervals.

Video Segmentation

In order to organize the video database, a segmentation algorithm is applied to the video data to obtain video intervals. These intervals may be organized into shot, group of shots, and story levels. The segmentation for shots is relatively simple and can be done automatically by shot-change detection algorithm based on low-level features. Gargi *et al* [124] studied a comparison of various video-shot-detection algorithms. These include: (1) color-histogram-based shot detection (CHSD) using different color spaces and similarity matrixes; (2) transform-domain algorithms on MPEG video using DCT coefficients; and

(3) block-matching methods for the optical-flow computation. These findings show that CHSD performed sufficiently well at a moderate computational cost, whereas MPEG-compressed domain shot detection algorithms are faster, but do not perform well as the CHSD.

Video segmentation at the higher levels can be more difficult as it requires high-level knowledge and depends on a particular domain application. Chang *et al* [123, 127] developed scene analysis and segmentation using an algorithm based on audio-visual information. The algorithm is a "computable" model framework in that the high-level scene segmentation can be reliably and *automatically* determined using low-level features. This work suggests two types of video scene: (1) N-type which is characterized by a long-term consistency of condition such as chromatic composition, lighting and sound; and (2) M-type (or MTV scenes) which is characterized by widely different visuals in terms of location, time and lighting conditions, assumed to be characterized by a long-term consistency in the audio track. Pfeiffer *et al* [137] used audio-visual information to detect scene changes. Here, similar shots are clustered according to the detected dialogue settings and similar audio content, in order to obtain the scene boundary.

While story segmentation is particular difficult due to its semantic meaning, upon taking advantage of high-level information in a particular domain application, story units can be segmented automatically [140–143]. In the TV news programs examined, Gao and Tang [141] used a combination of syntactic and semantic methods to identify anchorperson and new footage shots, in order to obtain segments of new stories that are interwoven with commercials. For a movie video, Hanjalic *et al* [142] suggest that a movie sequence can be broken up into "logical story units" which are characterized by overlapping links—the links that connect similar shots according to visual information and temporal variations. Thus, story boundaries can be detected at these points.

The challenge for this topic is in the segmentation of video according to object or region of interest [138, 139]. Zhong *et al* [138] utilized traditional region-based approaches to identify objects in the video, and incorporated user input for semantic object tracking. Indexing and retrieval can then be implemented at the object level. The object-based retrieval, however, is still undergoing. While identifying regions in the video may be easy, implementing and grouping them into meaningful objects is still done manually. Furthermore, object-based representation may not adequately deal with complex video content, i.e. it is difficult to refine the definite objects from the video.

Representation of Video Content

The traditional techniques use key-frames for video representation [134–136]. A few key frames are selected from a video shot for representing the content, and for similarity matching between shots. Jain *et al* [136] obtained a video clip by combining two or more shots, and used the sub-sampled frames for clip representation. In many cases, however, the key-frame representation is not sufficiently robust, since the selected key frames may not be effective representation of the other frames within the video. To overcome this problem, key-frame selection algorithms may be applied [135]. Here, frames are clustered according to a rate-distortion performance measure. Since video data has a high degree of frame-to-frame-correlation, a compact representation of the video can be achieved using the resulting clusters. However, while the KFVI method is relatively easy to implement, it produces a representation which may not be adequate to capture video content, since it does not take into account temporal information. Rather, the similarity matching between videos is based on the spatial content of the predefined key-frames. Furthermore, KFVI is not well adapted for representing video at scene and story levels.

Recently, content-analysis methods have focused on temporal information using motion features [122, 132, 133]. Jeong *et al* [133] introduced a fuzzy triple model to specify the relative spatial relationships between salient objects and their changes with respect to time. Ngo *et al* [132] employed a clustering technique to group together shots that are similar in motion and color features, and used the resulting clusters for retrieval. Fablet *et al* [122], used a non-parametric probabilistic method to capture dynamic content within shots, and applied the motion-related measurements to improve the motion features from the camera movement. According to the motion content, a hierarchical structure is built for a given database for indexing and retrieval of video shots. Sahouria *et al* [131] use principle-component analysis to select effective features for motion classification. This reduces complexity while improving effectiveness of classification. In general, using motion features for representing video content is useful for some parts of a video which have certain types of motions. However, within the motion class, there may be many sub-classes that cannot be well-separated by motion features.

High-Level Representation Techniques

Advances in pattern and speech recognition present new methods for the characterization and analysis of video content. The works in [121, 126, 130] demonstrate semantic indexing strategies that construct sets of *automatically* extracted key-words to describe video content.

Naphade and Huang [121] introduced a probabilistic framework for the semantic characterization of video content. A set of key-words is used to characterize video, based on the detected events (such as explosions or speech), scenes (indoors or outdoors), or objects (animal, planes). These attributes are referred to as multimedia objects, or *multijects*, that are detected automatically by low-level feature processing of the audio-visual data using machine learning methods, such as Hidden Markov Models (HMM) and Gassian Mixture Models (GMM). The pre-defined multijects can also be used to introduce other newly found multijects in the video, by adjusting the probabilistic quantity in a multiject network called a *multinet*. In this framework, indexing and retrieval are allowed by key-word method. Although challenging, this conceptual framework contains many of the aspects needed to be considered for practical implementation. However, the limited number of multijects used to define multimedia objects is not sufficient to exhaust video content, since many objects are difficult to detect. The number of defined multijects is also limited by the technology (e.g., recognition techniques). Furthermore, since the technique is based on a key-word search paradigm, some classes of object are more difficult to describe by key-words when compared to the visual content. In other words, the utilization of this framework is limited by the number of pre-defined multijects.

Upon the domain-specific application, event and object detection algorithms can be integrated to facilitate video browsing, filtering, and retrieval [126, 128]. High-level techniques such as human face detection, speaker identification, video caption recognition and event detection have been studied [126, 128, 129] for a particular type of video. These high-level techniques are, however, heavily dependent on specific attributes, such as particular events and objects, which makes extension for characterization of other unspecified contents very difficult.

2. Framework of Adaptive Video Indexing (AVI)

In the above discussion, two factors are required for effective video characterization: first, video representation should capture spatio-temporal information; second, there should be support for video indexing at various levels; that is, shot, group of shots, and story. While KFVI can be used at the shot level, temporal information is not addressed, hence it cannot be effectively used at the higher levels. On the other hand, the object, event, and caption detection algorithms, and automatic key-word indexing, are high-level methods which can be applied at the scene and story levels, but the limited number of 'terms' offered may not be adequate for characterizing the visual content within the video.

In view of this, a video representation based on a template-frequency model (TFM) is explored here to take into account spatio-temporal information. Video data is viewed as a collection of visual templates, so that the video characterization is the analysis of the probability of templates occurring in a video sequence. A visual template is regarded as a global view of objects, and other parts of the real-life images occurring in the video. The visual-template collection offers various "visual terms" to describe video content, and is more natural than a limited number of keywords used to annotate objects or events. Compared to the KFVI technique, that relies on a few representative frames and ignores other important parts using its key-frame selection algorithms, TFM differentiates degrees of importance among frames with weight parameters into a low-dimensional space, in order to fully explore temporal information at various levels and for differing video intervals.

Incorporating relevance feedback (RF) for improving retrieval accuracy is also important. While many RF models have been successfully developed for still-image application, they have not yet been widely implemented for video database application. The difficulty is that RF requires video representation to capture sequential information to allow analysis. So, while relevance feedback for video retrieval has been implemented, [45, 125], where the *audio-visual* information is utilized for characterizing spatio-temporal information within the video sequence, the application of RF to video files is, however, a time-consuming process, since users have to play each retrieved video file, which is usually large, in order to provide relevance feedback. In practice, this is a more difficult interaction with sample video files for retrieval on Internet databases. In the current work, the RF is considered an important method and will be implemented in automatic fashion. Since TFM representation emphasizes both spatial and temporal information, this allows RF to effectively analyze the dynamic content of the video. In section 5, TFM is integrated with a self-training neural network [43] to implement the automatic RF. This process further allows the improvement of retrieval accuracy, while minimizing user interactions.

3. Adaptive Video Indexing

The video indexing consists of three modules: video shot segmentation, template generation, and template-frequency modeling. In this section, video database characterization is first described, then followed by an explanation of the three modules.

Video Database Characterization

A video database is a collection of raw video streams. A single raw video stream is considered in the following definitions.

Definition 1. *A video stream F is a finite set of frames $f_1, f_2, ..., f_n$ that are ordered with respect to the index time n.*

Definition 2. *If $f_s, f_e \in F$ and $s < e$, then a video interval $I[f_s, f_e]$ over F is the set of frames $\{f_k \in F | s \leq k \leq e\}$.*

f_s and f_e of $I[f_s, f_e]$ are the starting frame and the ending frame, and they are denoted by $start(I)$ and $end(I)$, respectively. A video interval $I[f_s, f_e]$ is simply denoted by I, and $I(F)$ denotes the set of all intervals over F.

A video F can be organized into three levels: *shot, group,* and *story* levels. So, $I_{Shot}(F)$, $I_{Group}(F)$, and $I_{Story}(F)$, denote a set of all intervals over F, for shot, group, and story levels, respectively. A video group, then, is the stream of continuous shots having some contextual meaning. A video interval at the group level is defined by:

$$I_{Group} = \{I_{i,Shot} | \quad i = 1, 2, ..., S1, \} \quad I_{i,Shot} \in I_{Shot}(F) \tag{6.1}$$

Similarly, the story interval corresponds to:

$$I_{Story} = \{I_{i,Shot} | \quad i = 1, 2, ..., S2, \} \quad I_{i,Shot} \in I_{Shot}(F) \tag{6.2}$$

As story is the highest or most complex level, it usually contains a larger number of shots (i.e., $S1 < S2$).

Definition 3. *Let $\vec{x}_i \in \mathcal{R}^P$ represent a visual descriptor of frame f_i. A video interval $I[f_s, f_e]$ at any level is characterized by a set of video descriptors represented by $D_I = \{(\vec{x}_s, f_s), (\vec{x}_{s+1}, f_{s+1}), ..., (\vec{x}_e, f_e)\}$.*

D_I denotes a set of primary descriptors of I. It will be used for the video segmentation algorithm, and for obtaining a secondary descriptor used for the video indexing.

Intuitively, a *video descriptor database* VD for a video F is defined as a set of video descriptors for F and has the following form:

$$VD = \{(D_{I_1}, I_1), (D_{I_2}, I_2), ..., (D_{I_n}, I_n)\}, \tag{6.3}$$

Based on Eq.(6.3), video descriptor databases at the shot, group, and story levels are defined as follows:

$$VD_{Shot} = \{(D_{I_i}, I_i) | I_i \in I_{Shot}(F)\} \tag{6.4}$$

$$VD_{Group} = \{(D_{I_i}, I_i) | I_i \in I_{Group}(F)\} \tag{6.5}$$

$$VD_{Story} = \{(D_{I_i}, I_i) | I_i \in I_{Story}(F)\} \tag{6.6}$$

Note that D_I is regarded as the set of primary descriptors, and it is only used to characterize video at the frame level. In order to obtain video indexing, it will be reorganized into a higher level as a set of secondary descriptors (described in Section 3.0).

Shot boundary detection and feature extraction

In order to organize the video database, the video is first segmented into shots, which are the basic video unit. Then, the shots are joined into groups and stories according to chronological order within long videos. While shot segmentation is obtained automatically, the group and story level segmentations are manually obtained. Segmentation for video group and story is beyond the scope of this work, though an automatic process to achieve this may be found in [123].

The shot boundary is usually described by cuts, fades, wipes, or large camera motions. To detect shot boundaries, it is adopted here a color-histogram-based shot detection (CHSD) method which has proven to be more effective than any other, based on the comparison studies of [124].

Given a frame size of M-by-N, a frame difference measure is defined as:

$$F_D(t) = F_D(\vec{x}_t, \vec{x}_{t+1}) \tag{6.7}$$

$$= \frac{1}{\mathrm{M} \times \mathrm{N}} \sum_i |x_t[i] - x_{t+1}[i]| \tag{6.8}$$

where $\vec{x}_t$ denotes the color histogram for the t-th frame, t= 1,2,. . . ,T, and i indexes the histogram bins. The histograms computed on the H and S axes in *HSV* color space are used, where H and S are uniformly quantized into 16 and 3 regions respectively. This results in a 48-bin histogram. After applying the frame difference measure to video data, all frame pairs are mapped to a frame difference vector,

$$\vec{F}_D = \{F_D(t)|t = 1, 2, ..., T\} \tag{6.9}$$

Due to the significant change of contents at the shot boundaries, a cut is usually detected by applying a threshold to $\vec{F}_D$. This condition, however, is not effective when used to detect gradual transitions [124], because the frame differences in such transitions have a minimal change from one frame to the next, producing a sequence of small F_D value. To cope with this problem, and to generate evaluation values that more strongly reflect the transition, the frame difference measure is computed using two windows:

$$F_D(t) = F_D(\vec{x}_t, \vec{x}_{t+w_1}) + F_D(\vec{x}_t, \vec{x}_{t+w_2}) \tag{6.10}$$

where $w_1 = 1$ is applied for cut detection and $w_2 > 1$ is applied for dissolve detection. Experimentally, it is observed that a larger window size leads to better measured performance.

Template Generation

The template generation proceeds in two stages. In the first pass a training sample set is prepared from the entire video database. In the second stage, the visual templates $\vec{g}_r, r = 1, ..., R$ are determined and optimized by competitive learning [46, 85]. A color histogram resulting from the CHSD algorithm is used as a feature descriptor to characterize the corresponding video frames. The framework for video indexing, however, does not restrict us to a particular feature extraction scheme, and any available visual descriptor may be incorporated, since the goal is to measure the template-frequency quantity. Let a vector $\vec{x} = [x_1 ... x_{48}]^T \in \mathcal{R}^{48}$ be a color histogram feature vector. From the application of CHSD to all videos in the database, a feature matrix is obtained, $\mathbf{H}_v$,

$$\begin{aligned} \mathbf{H}_v &= [\vec{x}_1, ..., \vec{x}_v, ..., \vec{x}_V]^T \\ &= [x_{vi}],\ v = 1, ..., V,\ i = 1, ..., 48 \end{aligned} \tag{6.11}$$

where V is the total number of video frames. By randomly selecting subsamples in the matrix $\mathbf{H}_v$, a training set $\mathbf{H}_t$ is then obtained,

$$\mathbf{H}_t \subset \mathbf{H}_v \tag{6.12}$$

$$\mathbf{H}_t = [x_{ji}] \quad j = 1, ..., J, \quad J < V, \quad i = 1, ..., 48 \tag{6.13}$$

In order to convert all the values in the data set $\mathbf{H}_t$ to the same propositional scales, a scaling function $f(.)$ is applied to each column vector in $\mathbf{H}_t$, as follows:

$$f(\vec{x}_i) = \{\tilde{x}_{ji}| \quad j = 1, ..., J\} \tag{6.14}$$

$$\tilde{x}_{ji} = (x_{ji} - \bar{x}_{ji})/s_{ji} \tag{6.15}$$

$$\bar{x}_{ji} = \frac{1}{J}\sum_{j=1}^{J} x_{ji} \tag{6.16}$$

$$s_{ji} = \left(\frac{1}{J-1}\sum_{j=1}^{J}(x_{ji} - \bar{x}_{ji})^2\right)^{\frac{1}{2}} \tag{6.17}$$

where $\vec{x}_i = [x_{1i}, ..., x_{ji}, ..., x_{Ji}]^T$ is the i-th column vector of $\mathbf{H}_t$. $\tilde{\mathbf{H}}_t$ is assigned to denote a scaled version of $\mathbf{H}_t$. The scaling process (Eqs.(6.14) through (6.17)) assumes that J is large enough, such that $\bar{x}_{ji}$ and s_{ji} approximate the true mean and standard deviation of the distribution of all possible x_{ji}'s by the *Law of Large Number* (LLN) [8, 120]. This is to ensure that the scaling factors given by Eqs.(6.14)-(6.17) can be applied to new data that is not presented in the training set $\mathbf{H}_t$. This assumption is utilized for the mapping algorithm in Section 3.0.

After the scaling process, a subset of samples from the row vectors in $\tilde{\mathbf{H}}_t$ is associated to initialize the visual templates $\vec{g}_r, r = 1, ..., R$, where $R << J$. Let $\vec{x}_j \in \mathcal{R}^{48}$, $j \in [1, J]$ be a vector randomly selected from the row vectors in $\tilde{\mathbf{H}}_t$. Assuming that the current vector is assigned to template $\vec{g}_{r^*}$, i.e.,

$$||\vec{x}_j - \vec{g}_{r^*}|| < ||\vec{x}_j - \vec{g}_r||, \quad r = 1, ..., R, \quad r \neq r^* \tag{6.18}$$

using competitive learning, the value of $\vec{g}_r$ is updated as follows:

$$\vec{g}_{r^*}(n+1) = \vec{g}_{r^*}(n) + \ell(n)(\vec{x}_j - \vec{g}_{r^*}(n)) \tag{6.19}$$

The learning step size $\ell(n)$ is monotonically decreased according to the following linear schedule:

$$\ell(n+1) = \ell(0)(1 - \frac{n}{n_f}) \tag{6.20}$$

where n_f is the total number of iterations.

Template-Frequency Modeling

Within the AVI paradigm, a video datum is modeled using a set of visual templates, concatenated into a single feature vector through the assignment of numerical weights. This weighting scheme characterizes a differential degree of importance among each video frame, through a *template-frequency factor* specifying the occurrence of a particular visual template in the video, in the spirit of data modeling approaches described in [11, 44]. In other words, the template-frequency factor is used to estimate some statistical values which best identify the usefulness of the visual templates for describing the video contents, and aim at capturing spatio-temporal visual characteristics of the video content.

More precisely, each template vector $\vec{g}_r, r \in [1, R]$ is associated to a corresponding region $\Re_r \subset \mathcal{R}^{48}$ in a Voronoi Space (the Voronoi Space [85] is defined as the space containing the optimized-model vectors $\vec{g}_r, r = 1, ..., R$). Then, the contents of the entire video sequence (or interval) I is mapped to the regions, $\Re_r, r = 1, ..., R$ to obtain a weight vector $\vec{v} = [w_1, ..., w_r, ...w_R]$; each weight w_r specifies a statistic-measured value for the corresponding $\vec{g}_r$ according to *the degree of importance* of $\vec{g}_r$ to the video I.

Let $\vec{x}_m \in \mathcal{R}^{48}$ denote a vector extracted from the m-th frame in a set of descriptors $D_{I_j} = \{(\vec{x}_1, f_1), ..., (\vec{x}_m, f_m), ..., (\vec{x}_M, f_M)\}$, corresponding to an interval I_j. Given a set of visual templates $C = \{\vec{g}_r | \; r = 1, 2, ..., R\}$, a mapping of $\mathcal{R}^{48} \rightarrow C$ to a Voronoi Space is defined through:

$$\vec{x}_m \Rightarrow \langle \phi_v(C, \vec{x}_m), \quad \Re^{\eta}_{r*} \rangle \Rightarrow \rho^{(\vec{x}_m)} \tag{6.21}$$

$$\phi_v(C, \vec{x}_m) = \arg\min_r \left(||\vec{x}_m - \vec{g}_r|| \right), \tag{6.22}$$

$$\Re^{\eta}_{r*} = \cup^{\eta}_{i=1} \vec{g}_i \tag{6.23}$$

where $\Re^{\eta}_{r*}$ is a region containing η Voronoi cells neighboring $\vec{g}_{r*}$. Note that r^* denotes an index of the best-matched cell. By this definition, the given point $\vec{x}_m$ is mapped onto η cells (instead of one) neighboring to each other on the Voronoi space.

In Eq.(6.21), the region is specified from which samples are identified when a given visual template set is scanned. This region is data dependent in order to improve the embedding of correlation information. This is achieved through multiple-label indexing. For each vector $\vec{x}_m$, $m \in [1, M]$ the mapping proceeds in a number of distinct labels, which are specified as:

$$l^{\vec{x}_m}_{r*}, l^{\vec{x}_m}_{r*,1}, \ldots, \quad l^{\vec{x}_m}_{r*,(\eta-1)}. \tag{6.24}$$

Once a cell is selected, the η-1 neighbors which have not yet been visited in the scan are then also included in $\rho^{(\vec{x}_m)}$. This allows for interpretation of the correlation information between the selected cell and its neighbors. Since a video sequence usually has a very strong frame-to-frame correlation [123] due to the nature of time-sequence data, embedding correlation information through Eqs.(6.21)-(6.23) offers a better description for video contents, and thus a means for more accurate discriminant analysis. For example, two consecutive frames which are visually similar may not be mapped into the same cells; rather, they may be mapped onto two cells in a neigborhood area, so that mapping through multiple labels using Eq.(6.24) maps two frames from the same class in the visual space into the same class in feature space.

The visual content of the video frame f_m (associated with $\vec{x}_m$) is therefore characterized by the membership of $\rho^{(\vec{x}_m)}$. The resulting $\rho^{(\vec{x}_m)}$, $m = 1, ..., M$ of all frames from the mapping of the entire video interval I_j are concatenated into a single weight vector $\vec{v}_j$. This is represented through a weight scheme [44] by:

$$\vec{v}_j = (w_{j1}, ..., w_{jr}, ..., w_{jR}) \quad (6.25)$$

$$w_{jr} = \frac{freq_{jr}}{\max_r freq_{jr}} \times \log N/n_r \quad (6.26)$$

where the weight parameter $freq_{jr}$ stands for a raw frequency of template $\vec{g}_r$ in the video interval I_j (i.e., the number of times the template $\vec{g}_r$ is mentioned in the content of the video I_j); the maximum is computed over all templates mentioned in the content of the video I_j; N denotes the total number of videos in the system; and n_r denotes the number of videos in which the index template $\vec{g}_r$ appears. In this way, the weight w_{jr} balances two effects for clustering purposes: intra-clustering characterization and inter-clustering characterization. First, the intra-clustering similarity provides one measure of how well that template describes the video contents in the desired class, and it is quantified by measuring the raw frequency of a template $\vec{g}_r$ inside a video I_j. Second, the inter-clustering dissimilarity is quantified by measuring the inverse of the frequency of a template $\vec{g}_r$ among the videos in the collection, thereby specifying that the templates which appear in many videos are not very useful for the discriminant analysis.

It suffices here to note that only a few from the large number of templates is used for indexing an input video sequence (i.e., if the template $\vec{g}_r$ does not appear in the video I_j then $w_{jr} = 0$), so that the weight vector $\vec{v}_j$ is very sparse and only non-zero elements are kept.

Definition 5. *For a video descriptor database* $VD = \{(D_{I_1}, I_1), ..., (D_{I_j}, I_j), ..., (D_{I_J}, I_J)\}$, *where* $D_I = \{(\vec{x}_s, f_s), (\vec{x}_{s+1}, f_{s+1}), ..., (\vec{x}_e, f_e)\}$*,the indexing process produces a secondary video descriptor for each interval* I_j*, specified as* $\tilde{D}_{I_j} \equiv \vec{v}_j = (w^{\exists}_{j1}, ..., w^{\exists}_{jr}, ..., w^{\exists}_{jR})$*. The weight* w_{jr} *associated with a pair* $(\vec{g}_r, I_j)$ *is positive and non-binary.*

Adaptivity

Since the template-frequency model embeds all the visual contents occurring in a video sequence, the AVI technique can be adapted to characterize video sequences at different levels, from shot, group of shots, to story levels, ranking these in an ascending order of semantics. This allows for the system to facilitate the user's access to various levels as depicted in Fig. 6.1: (a) shot-to-shot, (b) shot-to-group, (c) group-to-group, (d) group-to-story, and (e) shot-to-story.

This architecture is able to accommodate retrieval from the lower to higher levels, e.g., retrieval of a video group or story by using a query from the shot or group levels. A user is generally seeking information across the different levels defined in the segmented videos. To satisfy this demand, it is expected that at a higher level, the video story should contain most of the visual contents occurring at the lower one. For instance, to retrieve a full news story, a small shot that contains the anchor can be utilized as a query.

Let I_{Group} be the video interval at the group level, which contains a total of $S1$ video shots, $I_{Group} = I_{1,Shot}, ..., I_{i,Shot}, ..., I_{S1,Shot}$ Then, a video descriptor for I_{Group} is given by

$$\tilde{D}_{I_{Group}} = (\vec{w}^{\exists}_{1,Group}, ..., \vec{w}^{\exists}_{r,Group}, ..., \vec{w}^{\exists}_{R,Group}) \quad (6.27)$$

where

$$\vec{w}_{r,Group} = \sum_{i=1}^{S1} w_{ir,Shot} \quad (6.28)$$

$\vec{w}_{ir,Shot}$ is the rth weight component of the ith shot $I_{i,Shot}$. Eqs.(6.27)-(6.28) are also applied to obtain a set of descriptors for a video interval at the story level, $\tilde{D}_{I_{Story}}$. A summary of the video description databases for each levels is as follows:

$$VD_{Shot} = \{(\tilde{D}_{I_{i,Shot}}, I_i) | I_i \in I_{Shot}(F)\} \quad (6.29)$$

$$VD_{Group} = \{(\tilde{D}_{I_{i,Group}}, I_i) | I_i \in I_{Group}(F)\} \quad (6.30)$$

$$VD_{Story} = \{(\tilde{D}_{I_{i,Story}}, I_i) | I_i \in I_{Story}(F)\} \tag{6.31}$$

In the querying process, a query can be chosen from VD_{Shot} or VD_{Group}, according to the links in Fig. 6.1. In these conditions, a query interval should follow two properties: first, it should be short enough to not contain many lengthy scenes. Second, it should be long enough that the context does not break down. This means that an interval from a story database may not be a suitable interval query.

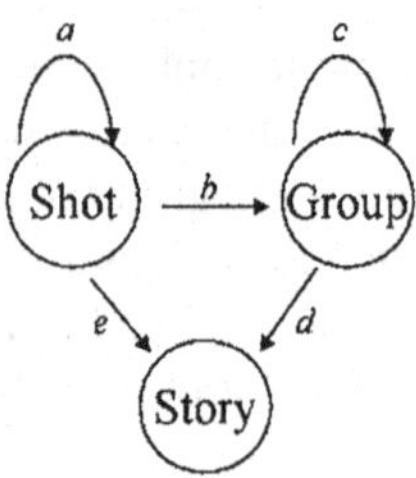

Figure 6.1. Multiple-level access to video database, *a*: shot-to-shot, *b*: shot-to-group, *c*: group-to-group, *d*: group-to-story, *e*: shot-to-story.

4. Application to CNN News Video Retrieval

This section describes an application of AVI for retrieval of CNN news videos. The performance of the AVI technique is studied by comparing it with the KFVI algorithm [41], which has become a popular benchmark for shot-based video retrieval. It is also demonstrates AVI performance in retrieving video groups and video stories, according to the links described in Fig. 6.1 (section 5.0).

Table 6.1 provides a summary of the test video data, obtained from the Informedia Digital Video Library Project [48]. This is the CNN broadcast news, which includes full news stories, news headlines, and commercial break sections. This video results in 844 video shots (see Fig. 6.2), segmented by the color histogram based shot boundary detection algorithm (described in section 3.0). A 48-bin histogram computed on HSV color space is used for both shot segmentation and indexing algorithms. The KFVI uses a histogram vector generated from a middle frame of the video shot as a representative video shot. The resulting feature database is scaled according to Eqs.(6.14)-(6.17). In the AVI case, a total of 5,000 templates are generated. Each video shot is described by its associated weight vector. This was generated by the template models, using neighborhood $\eta = 5$ [cf. Eq.(6.21)].

Video Sequences	# Sequences	# Cuts	# Frames	Lengths (min:sec)
Commercial	20	844	98,733	54:52
Movie clip	2			
Headline and story news	46			

Table 6.1. Description of sequences in the database: CNN broadcast news (at 352 resolution and 30 frames/sec.).

A total of 25 queries were made and the judgments on the relevance of each video to each query shot were evaluated. In general, the relevance judgment of videos is difficult because two video clips may be related in terms of the story context, and not just visual similarity. This fact was taken into account in this experiment, so a criterion employed here is a very subjective judgment of relevance: only retrieved video shots from the "same" stories were judged to be relevant. For example, four video shots shown in Fig. 6.3(a) were judged to be relevant because they were parts of the same stories. This also holds true for the video shots shown in Fig. 6.3(b).

Fig. 6.4 shows precision results as a function of top matches, averaged over all 25 queries. It can be observed that AVI performed substantially better than that of KFVI for every setting of the number of top matches (the average precision was higher by more than 18%). It is also observed that AVI is very effective in capturing spatio-temporal information from video, as seen in Fig. 6.5 which depicts retrieval results from the top sixteen best matches. It was observed that AVI allows similarity matching based on video contents, whereas the KFVI emphasis is on the content of the key-frame. There was a dominant brown color on the key-frame, degrading the performance of KFVI on this query.

Query by Video Clip

The purpose here is to demonstrate that the AVI technique can be adapted for retrieval beyond the shot level. According to the time line in the original unsegmented video, the shots are jointed into a meaningful group and story. Although there is an automatic technique available for detecting the news story [123], this has been done manually to ensure the quality of the segmented videos used for this experiment. Three feature databases were created to describe the videos in the three levels. The lengths of the video clips were between 0.5-43.5 seconds for the group level, and 5.7-180.3 seconds for the story level.

Figure 6.2. A subset of 844 video shots in the test database obtained from CNN broadcast news. For presentation purposes, each shot is shown by its corresponding key-frame.

In order to retrieve the video groups, six sets of video intervals were obtained for querying, $\{(I_{1,Shot}, I_{1,Group})_{q1}, ..., (I_{6,Shot}, I_{6,Group})_{q6}\}$, each of which was obtained from different stories. In the same set, the shot interval $I_{i,Shot}$ was one part of the group interval I_{Group_i}. This allows a comparison of the performance between query-by-video-shot and query-by-video-group. It is noted that lengths of the queries are as follows: $\{(1s, 1.9s)_{q1}, (2.1s, 3.3s)_{q2}, (2.4s, 12.3s)_{q3}, (15.3s, 39.3s)_{q4}, (2.8s, 4.5s)_{q5}, (1.3s, 3.5s)_{q6}\}$.

Figs. 6.6 (a)-(b) show the precision versus recall figures for all six sets of the test queries, resulting from the retrieval of the video groups. Fig. 6.6(c) shows a comparison between two links: shot-to-group (STG) and group-to-group (GTG). Evidently, the AVI exhibits a good accuracy for video group retrieval. We have an average precision of 90% at 50% recall, and more than 60% at 100% recall. It can be observed that the GTG link provides higher precision at lower recall levels, while the STG is superior at higher recall levels. This

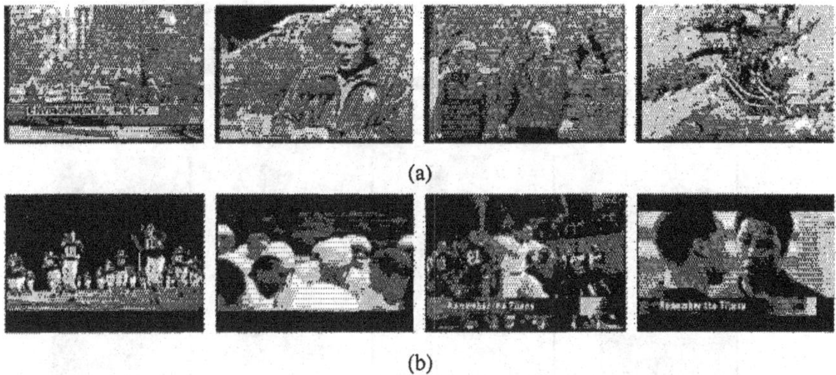

Figure 6.3. (a)-(b) Subjective judgment of relevance. In each case, the video shots are judged to be relevant to each other as they are originally from the same video story.

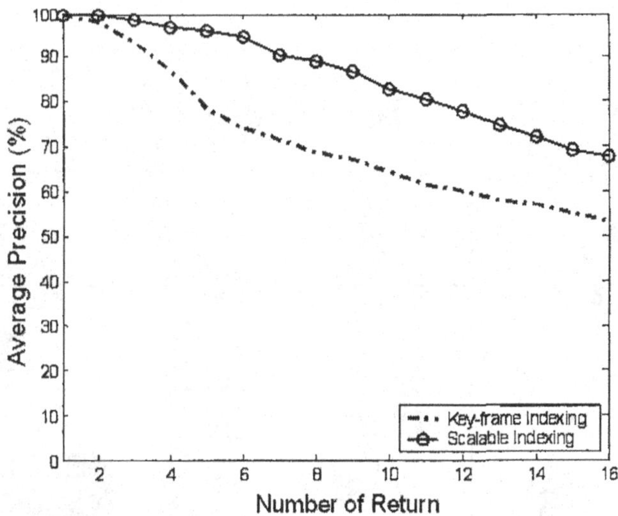

Figure 6.4. Average Precision Rate (%) as a function of top match, obtained during retrieval of 25 video shot queries.

is because video intervals at the group level usually contain more and longer information than at shot levels. On the other hand, a video shot usually contains less information, but can pinpoint relevance for ranking a video at higher recall levels.

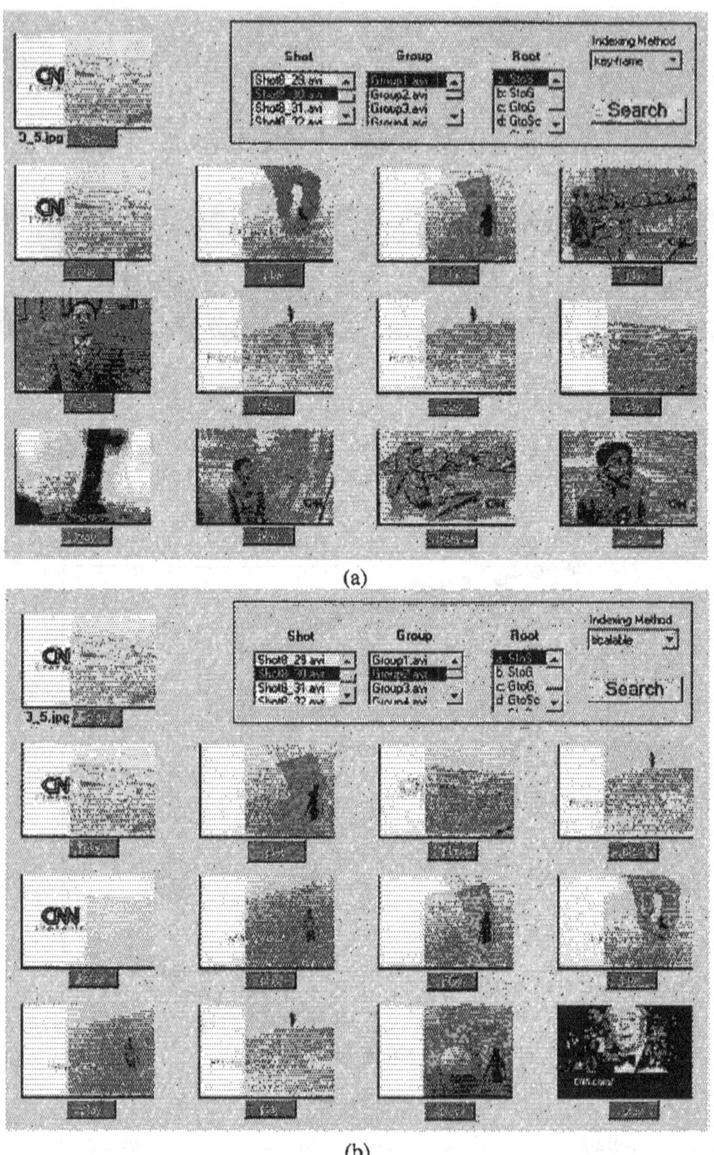

(b)

Figure 6.5. A comparison of the retrieval performance at the shot level; (a) obtained by KFVI; and (b) obtained by the AVI. In each case, the query is shown in the top-left corner, and the retrieved videos are ranked according to descending order of similarity scores, from left to right and top to bottom. It is observed that AVI allows similarity matching based on the video contents, whereas KFVI emphasizes the content of the key-frames effected by a color similarity.

Fig. 6.7 shows a group retrieval session, where a query clip contained two shots in a total length of 1.8 sec. For the sake of clarity, each of the retrieved clips is represented by a set of frames. It can be seen that the top 5 retrieved video clips are all relevant and are actually from the same story. A precise ranking of the relative similarity (to the query) among these retrievals may also be observed.

A possible application for retrieval of the video story is to utilize a news headline to retrieve the full news story. This enables one user to go directly to the full story from the headline of interest. Five news stories introduced with at least two headlines (summarized in Table 6.2) are examined. Then five shots and five video groups from the news headline are utilized for the test. Fig. 6.8 shows the AVI performance in retrieving the news stories by employing the shots and groups from news headlines. It is observed that all relevant video segments related to the same story were retrieved at close to 50% precision. This means that, on average, all relevant video intervals can be retrieved within the first ten retrievals. A comparison of STS and GTS performances shows that STS performed slightly better than GTS at lower recall levels, while GTS is much better at higher recall levels [cf. Fig. 6.8]. This is opposite to the results from retrieving video groups. In this case, a great deal of information from the video group is favorable for retrieving the video story.

Query	# Story	Result: (recall, precision)	
		Shot-to-Story	Group-to-Story
Query 1	S7, S22, S31	(0.33, 1.00), (0.67, 0.50), (1.00, 0.50)	(0.33, 1), (0.67, 0.50), (1.00, 0.50)
Query 2	S5, S8, S14, S28, S29	(0.20, 1.00), (0.40, 0.67), (0.60, 0.50), (0.80, 0.11), (1.00, 0.10)	(0.20, 1.00), (0.40, 0.50), (0.60, 0.50), (0.80, 0.50), (1.00, 0.23)
Query 3	S1, S26, S46	(0.33, 1.00), (0.67, 1.00), (1.00, 0.75)	(0.33, 1.00), (0.67, 1.00), (1.00, 1.00)
Query 4	S36, S37, S50, S55	(0.25, 1.00), (0.50, 1.00), (0.75, 1.00), (1.00, 0.50)	(0.25, 1.00), (0.50, 1.00), (0.75, 0.38), (1.00, 0.31)
Query 5	S7, S36, S39	(0.33, 1.00), (0.67, 0.40), (1.00, 0.50)	(0.33, 1.00), (0.67, 0.52), (1.00, 0.43)

Table 6.2. Story retrieval results, comparison between two links: shot-to-story (STS) and group-to-story (GTS).

Fig. 6.9 shows an example of story retrieval. The query clip [cf. query 2 in Table 6.2] is a news quiz about President G. W. Bush visiting a national park. There are five relevant stories in the database. The results show that

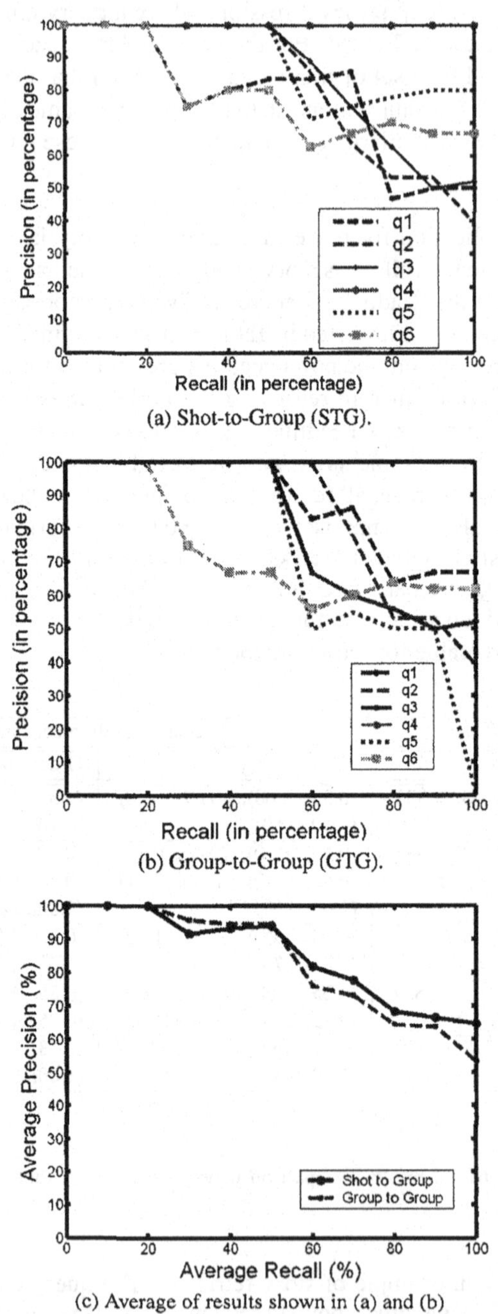

(a) Shot-to-Group (STG).

(b) Group-to-Group (GTG).

(c) Average of results shown in (a) and (b)

Figure 6.6. Precision and recall rates of retrieval of the video groups, using six query sets, employing the two links: (a) Shot-to-Group (STG) and (b) Group-to-Group (GTG). The average precision of (a) and (b) are shown in (c).

Figure 6.7. Top 5 retrievals ((b)-(f)) answering to a query clip that contains two video shots ((a)), using the group-to-group link. Note that the first ranked video is the query itself. All of the retrieved clips are originally from the same story.

relevant segments were retrieved at rank 1, 4, 6, 8 and 22. It is observed that the performance degradation on video rank 22 is a result of containing many irrelevant shots—this video segment summarized the news.

5. Automatic Reclusive Video Retrieval

Many years of research on *interactive* retrieval have been focused on image database applications [8, 11, 55, 56, 61], while only limited progress has been made for video database applications [45, 166]. In contemporary interactive systems, the challenging issue is to reduce user workload; the systems aim at obtaining accurate retrieval results with minimum user input—with fewer

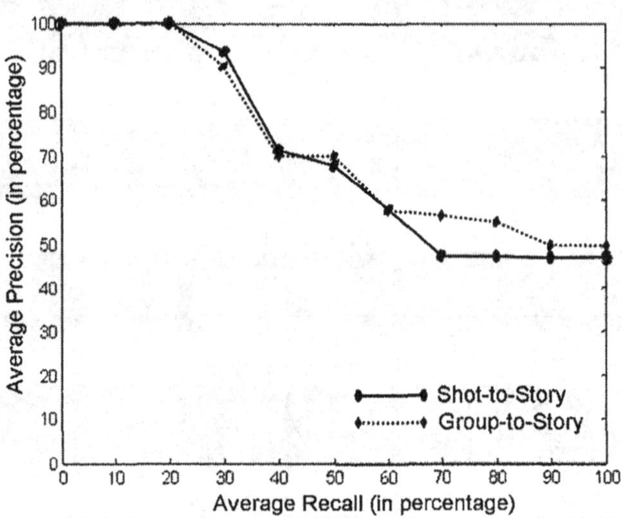

Figure 6.8. Precision and recall rates of retrieval of the video stories, employing the two links: Shot-to-Story (STS) and Group-to-Story (GTS).

feedback cycles, and a smaller set of training samples. These requirements are particularly essential for the application to video data stored in a networked database, with many constraints such as time and bandwidth.

Furthermore, interactive systems require a *self-adaptation* process to achieve high retrieval performance and minimize user input. Since feedback samples are usually required for the adaptation of relevance feedback systems, it is difficult to apply many cycles of the relevance feedback to network-based video databases, considering the transmission bandwidth and time constraints. Under the traditional relevance feedback paradigm, retrieval ability is entirely based on the amount of feedback samples [45] and the ability of the user to provide consistent continuous feedback. *Automatic relevance feedback* can be more flexible than this process. In particular, the information the user would traditionally provide through relevance feedback interfaces is not longer required. This process now involves the "automatic" modification of self-organizing neuronal weights to construct relevance feedback parameters.

In this section, this automatic strategy is further developed with adaptive video indexing (AVI) and self-adaptation methods, incorporating these into a high performance semi-automatic relevance feedback. AVI indexing structure

(a) query

(b) Rank 1: Story 5, clip length: 35.70 seconds

(c) Rank 4: Story 14, clip length: 36.30 seconds

(d) Rank 6: Story 8, clip length: 2 minutes and 34.9 seconds

Figure 6.9. Story retrieval using group-to-story link; (a) query clip containing two video shots (3.7 sec). There are a total of five relevant stories in the database, showing President G.W. Bush visiting a national park: (b) the first segment (35.7 sec) is ranked at no. 1; (c) the second segment (36.3 sec) is ranked at no. 4; (d) the third segment (34.9 sec) is ranked at no 6; (e) the fourth segment (30.4 sec) is ranked no. 8; (f) the fifth segment (20.5 sec) is ranked at no. 22.

is very effective in implementing adaptive systems in both user-controlled and fully automatic modes. In these two mode operations, it will be demonstrated that the effectiveness of relevance feedback analysis will rely on accurate modeling of spatio-temporal information. The chapter will demonstrate the appli-

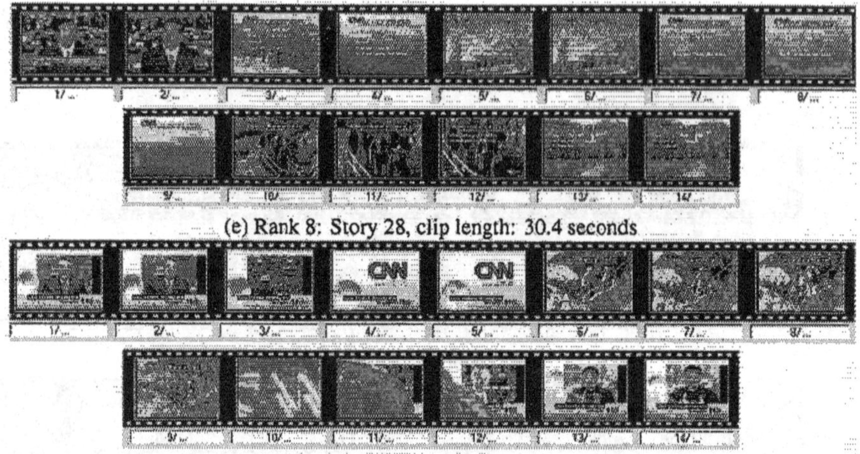

(e) Rank 8: Story 28, clip length: 30.4 seconds

(f) Rank 22: Story 29, clip length: 20.5 seconds

Figure 6.10. (*cont.*) Story retrieval using group-to-story link; (a) query clip containing two video shots (3.7 sec). There are a total of five relevant stories in the database, showing President G.W. Bush visiting a national park: (b) the first segment (35.7 sec) is ranked at no. 1; (c) the second segment (36.3 sec) is ranked at no. 4; (d) the third segment (34.9 sec) is ranked at no 6; (e) the fourth segment (30.4 sec) is ranked no. 8; (f) the fifth segment (20.5 sec) is ranked at no. 22.

cation of AVI by implementing the relevance feedback in an automatic fashion. In this paradigm, the adoption of the adaptive signal propagation network [43] is proposed to implement the automatic relevance feedback. Since neural network models perform effectively when matching given patterns against a large number of possible templates, this organization can be adopted for similarity matching in video retrieval. The AVI parameters are associated to the network weights to reorganize the parameters through a signal propagation process within the network. This process improves retrieval accuracy, while minimizing user interaction.

The Automatic Relevance Feedback Network (ARFN)

As it is observed in section 3, AVI models a video by using numerical weight parameters, $w_r, r = 1, ..., R$, each of which characterizes a *degree of importance* of the visual templates presented in the video. These weight parameters will be re-organized on a per query basis. At this point, video clusters that maximize the similarity within a cluster, while also maximizing the separation from other clusters, can be formed based on content identifiers, to *initialize* the ranking for answering a query. This ranking is now adopted to re-organize the degree of importance of the visual templates through the following process. First, the process identifies "effective templates" that are the common templates among videos in a retrieved set. Then, those templates considered to be the most significant for reweighting the existing templates of the initially submitted query are weighted, to improve the performance of ranking. In other words, we allow the templates that are not presented by the initially submitted query (i.e., $w_r = 0, r \in [1, R]$), but are common among the top-ranked videos (i.e., the potentially relevant videos), to "expand". This results in reorganization of the degree of importance of the query's templates for better video similarity measuring.

This process is in the same spirit as the user-controlled relevance feedback techniques widely used in information retrieval applications. More specifically, a set of significant *terms* defined of the items specified by the user is added to the initial query, and used to reweight the query components [11, 44]. In this work, we similarly adopt this query reformulation scheme for the expanding of queries to improve ranking. However, our goal here is to minimize user involvement, by proposing the adoption of a neural network model [43]. As neural networks can perform very well at matching a given pattern against a large number of possible templates, we use this structure for selecting relevant videos. Fig. 6.11 shows the neural network architecture for automatic video ranking.

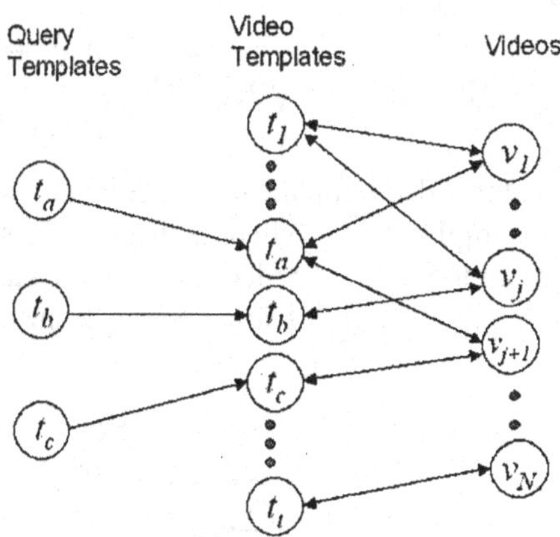

Figure 6.11. A neural network architecture for automatic relevance feedback.

A General Model

The network is composed of three layers: one for the query templates, one for the video templates, and the third for the videos themselves. Each node has a connection weight communicated to its neighbors via the connection links. The query template nodes initiate the inference process by sending signals to the video template nodes. The video template nodes then themselves generate signals to the video nodes. Upon receiving this stimulus, the video nodes, in their turn, generate new signals directed back to the video template nodes. This process might repeat itself several times, through the second and the third layers, which allows the network to find templates that appear to be relevant on the basis of initial ranking, and use those templates to refine the ranking process.

To be precise, let $\vec{v}_j = \{w_{qr} | r = 1, ..., R\}$ denote the set of the query's weight components, obtained by converting the video query v_q into a set of templates. Let $mesg_{r^{(q)} \to r^{(t)}}$ denote the message sent along the connection $\{r^{(q)}, r^{(t)}\}$ from the r-th query node to the r-th video template node. Also, let $mesg_{r^{(t)} \to j^{(v)}}$ denote the message sent along the connection $\{r^{(t)}, j^{(v)}\}$ from the r-th video template node to the j-th video node, $j \in [1, N]$. Note that $mesg_{r^{(q)} \to r^{(t)}}$ is a one-to-one correspondence, while $mesg_{r^{(t)} \to j^{(v)}}$ is a one-to-many correspondence. First, each query template node is assigned a fixed activation level equal to $a_r^{(q)} = 1, r \in [1, R]$. Then, its signal to the video template node is attenuated by normalized query template weights $\bar{w}_{qr}$, as follows:

$$mesg_{r^{(q)} \to r^{(t)}} = a_r^{(q)} \times \bar{w}_{qr} \tag{6.32}$$

$$\bar{w}_{qr} = \begin{cases} \frac{w_{qr}}{\sqrt{\sum_{r=1}^{R} w_{qr}^2}} & \text{if } \vec{g}_r \in v_q \\ 0 & \text{otherwise} \end{cases} \tag{6.33}$$

When a signal reaches the video template nodes, only the video template nodes connected to the query template nodes will be activated. These nodes might send new signals out, directed towards the video nodes, which are again attenuated by normalized video template weights $\bar{w}_{jr}$ derived from the weights w_{jr}, as follows:

$$mesg_{r^{(t)} \to j^{(v)}} = mesg_{r^{(q)} \to r^{(t)}} \times \bar{w}_{jr} \tag{6.34}$$

$$\bar{w}_{jr} = \begin{cases} \frac{w_{jr}}{\sqrt{\sum_{r=1}^{R} w_{jr}^2}} & \text{if } \vec{g}_r \in v_j \\ 0 & \text{otherwise} \end{cases} \tag{6.35}$$

Once the signals reach a video node, the activation level of this video node (associated to the video v_j) is given by a sum of the signals (the standard cosine measure),

$$a_j^{(v)} = \sum_{r=1}^{R} mesg_{r^{(t)} \to j^{(v)}} \tag{6.36}$$

$$= \sum_{r=1}^{R} \bar{w}_{qr}\bar{w}_{jr} = \frac{\sum_{r=1}^{R} w_{qr}w_{jr}}{\sqrt{\sum_{r=1}^{R} w_{qr}^2}\sqrt{\sum_{r=1}^{R} w_{jr}^2}} \tag{6.37}$$

This finishes the first round of signal propagation. The network output (i.e., $a_j^{(v)}$, $j = 1, ..., N$) is a desired ranking of the videos for retrieval. The process, however, does not stop here. The network continues the ever-spreading activation process after the first round of propagation. This time, however, a minimum activation threshold is defined such that the video nodes below this threshold send no signals out. Thus, the activation level at the r-th video template node is obtained by summing up the inputs from the *activating* video nodes as follows:

$$a_r^{(t)} = \sum_{j \in Pos} a_j^{(v)} \bar{w}_{jr} \tag{6.38}$$

where $a_j^{(v)}$ denotes the activation levels of the j-th video node and *Pos* is the set of j's such that $a_j^{(v)} > \tau$, where τ is a threshold value. The activation process is allowed to continue flowing forwards and backwards between the video template nodes and the video nodes, inducing an order to the videos, based on the corresponding node activations at each stage.

In other words, we allow the network to automatically expand the query templates analogous to the relevance feedback model [11, 44]. The signal propagation process is directly related to the derivation of new weights of query templates, whereby a new template appearing in the highly activated videos, despite not having appeared in the original query, may become active and activate other videos. This modifies the initial vector ranking in the retrieval process.

Fig. 6.12 graphically describes the spreading activation process. Fig. 6.12(a) shows two query templates sending signals to the video template nodes $\{a, b\}$. The video nodes: $\{c, d, e\}$ are activated (the application of the threshold is omitted to simplify illustration). Fig. 6.12(b) shows the signals propagating backward to the video template layers. At this time, f, g, and h are the newly activated nodes. After re-calculating the node activations, the video template nodes send signals forward to the video nodes as shown in Fig. 6.12(c). This results in a new ranking, which includes a new video node, i. We see that the network then utilizes video template node f, present in the initial ranking, to find more relevant video nodes.

A Modified Model

A new activation level computed in Eq.(6.38) can be viewed as a modified weight of the query template, where only videos with significant activation levels are considered to be good candidates for modifying the query template activations. Practically, anti-reinforcement learning is adopted to improve speed of convergence. [11, 44], whereby both original query components and a negative feedback strategy can help to improve effectiveness. Thus, as an alternative to Eq.(6.38) we derive the following formula for the activation of the r-th video template node:

$$a_r^{(t)} = \frac{l_r}{\left(\sum_{r=1}^{R} l_r^2\right)^{1/2}} \tag{6.39}$$

$$l_r = w_{qr} + \alpha \sum_{j \in Pos} a_j^{(v)} \bar{w}_{jr} + \beta \sum_{j \in Neg} a_j^{(v)} \bar{w}_{jr} \tag{6.40}$$

where $a_j^{(v)}$ is the activation level of the j-th video, *Pos* is the set of j's such that $a_j^{(v)} > \tau$, and *Neg* is the set of j's such that $a_j^{(v)} < -\tau$, where τ is a threshold value. In addition, α and β are the suitable positive and negative constant values. In the experiments reported in Section 5.0, the system performance was based on $\alpha = 0.95$ and $\beta = -0.5$.

Retrieval of Video Short with ARFN

Next ARFN is applied to improve retrieval accuracy for the CNN News database discussed in section 3, which contains 844 video shots. This video database results in a network with 5,844 nodes and 14,800 connections. The results of three tests are showed: letting the activation spread for one, three, and twenty iterations. The parameters were set at $\tau = 0.1$, $\alpha = 0.95$ and $\beta = 0.05$ [cf. Eq.(6.40)]. Table 6.3 shows the improvement of the average precision in retrieving 25 queries.

The following observations were obtained from the results. Firstly, the ARFN was very effective in improving retrieval performance—the average precision increased by more than 11%, and is particularly significant in the top 10 to 16 retrievals. Secondly, it is stabilized very quickly. Thirdly, allowing many

No. of Return	Cosine measure	1 Iter.	3 Iter.	20 Iter.	User-controlled RF
1	100	0.0	0.0	0.0	0.0
2	100	0.0	0.0	0.0	0.0
3	98.67	+1.33	+1.33	-1.33	+1.33
4	97.00	+1.00	+2.00	-2.00	+3.00
5	96.00	+0.80	+1.60	-1.60	+4.00
6	94.67	+0.67	+2.00	-1.33	+5.33
7	90.29	+2.29	+3.43	+1.14	+9.71
8	89.00	+3.00	+2.50	+1.50	+11.00
9	86.67	+2.67	+3.11	+2.67	+12.89
10	82.80	+5.60	+5.60	+4.80	+16.00
11	80.36	+6.18	+6.18	+5.09	+17.82
12	77.67	+7.33	+7.67	+7.00	+18.33
13	74.77	+8.62	+10.15	+8.31	+18.77
14	72.00	+9.43	+11.14	+9.72	+20.29
15	69.33	+9.87	+11.20	+10.13	+21.87
16	67.75	+9.50	+11.00	+10.00	+20.75

Table 6.3. Average Precision Rate, APR (%) obtained by automatic relevance feedback network (ARFN), using 25 video shot queries. ARFN results are quoted relative to the APR observed with cosine measure.

iterations degraded the performance slightly. Finally and most significantly, the results were achieved by simply allowing activation flow, with no user input.

It was observed that the values for τ, α and β affected the results and confirmed the reports of other studies [11, 27, 44] with regard to the value for α. However, the identification of the proper values for these parameters was completed conveniently as they were usually found in certain ranges. It was also observed that without applying the threshold level τ, it was found only modest improvement initially, while all nodes became increasingly activated. This leads to a longer processing time and to random ordering of the videos.

Based on the same setting of parameters α and β, we obtained retrieval results by user-controlled relevance feedback for the query modification model in Eq.(6.40) [44]. Here, user provided relevance feedback on each retrieved video from the top-16 best matches, and the results obtained are shown in the last column of table 6.3. It was observed that the user-controlled relevance feedback gave better performance than that of the ARFN: on average 7.1% higher. However, it should be taken into account that the users had to provide feedback on each of the videos returned by a query in order to obtain these results.

6. Chapter Summary

Video database applications require a suitable indexing technique to capture the time-varying nature of video data, together with a high performance retrieval strategy. This chapter presented an adaptive video indexing (AVI) technique and its integration with the specialized neural network model, which impressively satisfies these requirements. The template-frequency model (TFM) demonstrates very effective representation of the dynamic content of video, at various levels, and offers retrieval-by-video-clip to facilitate retrieving video groups and stories. Since TFM model addresses the difficulty in capturing spatio-temporal information, it allows relevance feedback analysis to capture information on the dynamic content, rather than spatial information offered by key-frame based modeling techniques. Unlike previous video database search attempts, we incorporated a self-learning neural network to implement an automatic relevance feedback method, which requires no user input for its adaptation. Based on a simulation study, this adaptive system, utilizing the TFM and automatic-RF retrieval architecture, can be effectively applied to a video database, with promising results. This approach combines many new features, which may help to usher in a new generation of video database applications.

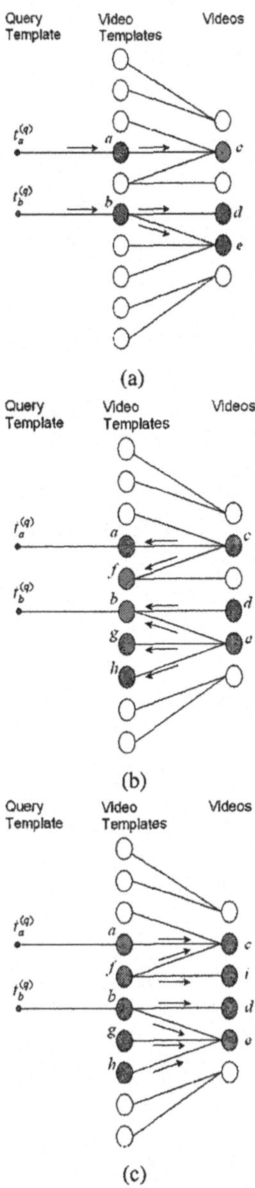

Figure 6.12. Signal propagation: (a) signals from the two query template are sent to the video template nodes, and three video nodes {c, d, e} are activated; (b) the signal propagates back from the third layer to the second layer, resulting in more activated video temple nodes; (c) signal propagates back to the third layer. This results in the activation of new video nodes by expanding the original query template and the activated video nodes in (b).

Chapter 7

MOVIE RETRIEVAL USING AUDIO-VISUAL FUSION

1. Introduction

This chapter presents applications of AVI techniques for retrieval of movies. The first part will discuss an online system that uses AVI to implement interactive search and retrieval of movies on the Web. The system will demonstrate the AVI technique in relevance feedback mechanism controlled by user as well as self-organizing learning units, and compare retrieval performance (both accuracy and processing speed) to other popular retrieval techniques. The second part of the chapter will study an audiovisual fusion model for retrieval of movies in replying to concept-based queries. AVI is applied to characterize visual content which is then combined with audio content descriptor using a fusion model. This model also integrates relevance feedback to achieve a high performance retrieval.

2. Interactive System for Movie Retrieval

Human-centered search and retrieval have been extremely central to multimedia computing applications. Contemporary multimedia systems, such as the MARS [8] and PicSom [61], were developed recently to overcome the obstacles of this challenging issue. In these practical applications retrieval accuracy is not the only concern. A user friendly environment is also highly desirable. Ideally, human-machine interaction in multimedia search engines will have a small set of training samples (typically, less than 20 video samples), and minimum user input (one to two cycles of relevance feedback). To that end, it is presented here as a web-based interactive video retrieval system, Interactive-based Analysis and Retrieval of Multimedia (iARM), which was successfully applied to the human-centered video retrieval application and can be accessed online [144].

The iARM system is a web-based search engine, and has been implemented using AVI video indexing structure and an interactive content-based retrieval strategy. This section demonstrates its application for retrieval of "video clips" from a general-purpose video database of 20 hours, which includes 14 Hollywood movies segmented into a total of 2,401 video clips, each of which is approximately 30 seconds long. Table 7.1 gives the details of the database. This system uses no textual information in the matching process to explore the possible advances of CBVR. In a real-world application, however, textual information is often used as a helpful addition to CBVR systems, in particular to provide preliminary information concerning video files for query interface as well as retrieval.

The iARM system is implemented to manage a centralized database using the Java 2 Enterprise Editions (J2EE), shown in Figure 7.1. In this system, the video database is located on the single server, and the system provides user interactions through Java Server Page (JSP) interface. The requests and feedback on the client sides are implemented through JSP, which are then processed within the Java Bean on the server side. The current implementation of the iARM system provides only two query interfaces: a video ID search and a query-by-example method. In the later phase, it will include several more query interfaces methods, such as a textual interface, a JAVA-based drawing, and a CGI-based Web interface for submitting a query video of any format anywhere on the Internet.

Performance on Video Queries

The AVI technique described in Chapter 6 was applied for indexing all video clips in the database. In a separate off-line process, each video clip was indexed by a set of visual templates. These were generated by the competitive learning algorithm; a training set of 16,681 samples was created by sub-sampling video frames in the database, and then it was optimized to generate 2,000 templates. During the online process, the video search is initiated by the query submitted by the user and followed by relevance feedback. Using the same strategy as the relevance feedback technique described in [44], a new query was obtained by enhancing the relevant models and suppressing the irrelevant models from the original query (cf. Eq. 2.3). The objective at this point was to evaluate the AVI in its capability in modeling human perception in the interactive retrieval system, so that automatic relevance feedback was not considered in this part of the experiments. However, the automatic results will be discussed in Section 2.0 with the semi-automatic process.

To provide numerical results, the system was tested by 40 sample video clips chosen randomly from fourteen movies. For each query, relevance feedback was provided on the top 16 best match samples, and the precision was recorded. Table 7.2 shows the retrieval results averaged over the 40 queries. From the table, it is observed that the iARM system gave a very high precision of 73.6% at the initial stage (i.e., more than 11 relevant video clips were retrieved out of the top 16 best matches). It was also observed a significant improvement of 90% precision was achieved after a single feedback cycle. This implies that the AVI method is highly effective in capturing spatio-temporal information from video. This also indicates that AVI is efficient and highly adaptable, as only a single user feedback is required for significant improvement.

Table 7.2 also shows the comparison between AVI and other video indexing methods that use video clustering strategies for the video content characterization. The compared methods are denoted as KFVI (key-frame based video indexing). Here, the KFVI employed video clustering approaches discussed in [135, 132, 164] for selection of the representative frames from video clips. In this way, for each video clip, frames are clustered based on frame descriptors, and frames that are close to the cluster centriods are selected as key frames. The k-mean algorithm and clustering validity method demonstrated in [132] were employed for selection of key frames.

KFVI process started by extracting 48-bin color histogram vector from each video frame in a given video clip. Then, it applied the clustering algorithm to the resulting histogram vectors to make a k-means with different values of k, for $k = \{1, 2, ..., 10\}$. The k-means is run multiple times for each k, and the best of these is selected based on sum of squared errors. Finally, Davies-Bonldin index [167] was calculated for each k, $k \in \{1, 2, ..., 10\}$, and the k that gave the smallest Davies-Bonldin index was chosen. In doing so, the optimum number of cluster will vary according to the cluster validity analysis of the resulting clusters. The closest frames to the clusters (one frame from each cluster) were selected as key frames, and this method is denoted as MKF (multiple-key frame) method. For comparison, a single frame which is the closest frame to a cluster centriod was selected as a key frame, and this method is denoted as SKF (single key frame) method. Video content similarity matching used by the SKF was obtained by comparing the descriptor vectors of the selected key frames of the query and the target videos. However, for the MKF method, similarity measure was obtained by matching multiple key frames of query against multiple key frames in the target video clips. To be precise, let S be a similarity score. The similarity was obtained by: $S = \sum_{i=1}^{N} s_i | \; s_i = \min_{j=1,\ldots,M} \{d[i,j]\}$, where $d[i,j]$ is the distance between the i-th key-frame of query and the j-th key-frame of the target video; N and M are the total number of key-frame of the query and target videos, respectively.

From the results, it was observed that although the SKF method can be used for retrieval of video shots, SKF is less effective in characterizing video content of video clips. The SKF result was at 39.22% precision. By considering multiple key frames as in the MKF method, the performance of the key-frame based video indexing method can be improved to 62.34% . However, this result is approximately 10% less precise than that of AVI.

Automatic and Semi-Automatic Retrieval

In order to achieve a high retrieval performance, iARM system was implemented by the automatic and semi-automatic retrieval algorithms employing ARFN architecture. In this case, depending on the network traffic conditions, users can submit automatic and semi-automatic queries, thereby the automatic query can avoid the transmission of training sample video files over the network. Using the same set of queries as in the previous results, this system first performed an automatic retrieval for each query to adaptively improve its performance. After three iterations of signal propagation in the ARFN, the system was then assisted by the users. Table 7.2 provides the summary of the retrieval results, obtained by automatic and semiautomatic methods. It was observed that the semiautomatic method is superior to the automatic method and the user interaction method. The best performance was achieved at 92.03% precision. In addition, the moderate performance of the automatic method can be beneficial to the user when the network resource is limited.

Robustness

The strength of the iARM system was evaluated against a variety of templates used by AVI for indexing video clips. Specifically, three sets of templates at R =500, 1000, and 1500, were generated using a competitive learning algorithm, where R denotes the number of templates. These are approximately 3%, 6%, and 9% of the training sample set, respectively. For each set of templates, video files were indexed, and the resulting feature databases were used by the retrieval system. Table 7.3 shows the retrieval results obtained by retrieving 40 queries. It was observed that the iARM system is fairly robust compared to the deviation of number of template R. The results in the second column show the small variation in retrieval performance (i.e., 1.09% reductions of precision) as the number of templates was reduced from 2000 to 1000 templates. In the aforementioned case, video matching used the non-adaptive matrix. However, the performance gap is increased to 2.34% after relevance feedback as we had anticipated, since human interpretation has been addressed to some extent by the interactive algorithm. The results show that a set of templates with $R \geq 1000$ is sufficient to model the video content in this database. This value of R is approximately 0.1 percent of the entire video data in this collection.

Speed

The complexity of all algorithms presented in this system can be divided into two parts. The first is the off-line process which includes (1) template generation using competitive learning algorithm, and (2) video indexing using AVI. The second part is the online process for query execution which starts from the time the user submit a query to the time the system finishes returning a list of the retrieved videos. Although, the complexity of the two processes needs to be considered in practical implementation, the most important aspect is the online process for executing a query in realtime, which needs to be completed rapidly.

The algorithms in the offline process have been implemented on a Pentium IV 2.4 GHz PC using the Window XP system and MATLAB implementation, whereas the online process used a Pentium IV 2.0 GHz PC Server system and JAVA implementation. Table 7.4 summarizes the processing times used by all algorithms on different sets of templates. As reported in the table, to compute a feature vector associated with a single video clip requires approximately 6 seconds, at $R = 2000$. The speed is much faster when the system implemented a smaller set of templates, i.e, at $R = 500$. However, as shown in the last column of the table, the matching speed is considerably fast regardless of the value of R. This system took on average less than 0.7 seconds of CPU time to complete a single retrieval process. Note that the query execution time is averaged because it was dependent on the available resources within the server at the time it is measured.

3. Audio-Visual Cue

Previously, the application of AVI discussed concerns only with visual information of content-based video retrieval. However, it is known that the spatio-

ID	Movie Name	ID	Movie Name
1	15 Minutes	8	Lion King
2	40 Days & 40 Nights	9	US Marshals
3	A Beautiful Mind	10	Me, Myself & Irene
4	Dr. T	11	The Adventures of Pluto Nash
5	Final Fantasy	12	Romeo Must Die
6	Gladiator	13	Scooby-Doo
7	Just Visiting	14	The Two Towers

Table 7.1. Detail of test video database, containing 20 hours of Hollywood movies.

Methods	Average Precision (%)	Methods	Average Precision (%)
Initial Result without RF	73.59	KFVI(SKF)	39.22
User Feedback	90.47	KFVI(MKF)	62.34
Automatic RFN	80.12		
Semi-Automatic Retrieval	92.03		

Table 7.2. Average precision (%), obtained by retrieveing 40 queries, measure from the top sixteen retrievals, using user-controlled relevance feedback, automatic relevance feedback, semi-automatic retrieval methods.

# Models	Average Precision (%)	
	Initial Result	After 1 feedback
R=2000	73.59	90.47
R=1500	72.81	89.69
R=1000	72.50	88.13
R=500	68.75	81.25

Table 7.3. Average precision (%), obtained by retrieving 40 queries, when using four sets of models for video indexing.

temporal information of video files contains a fusion of multimodality signal of visual frame, music, caption text, and spoken words that can be deciphered from the audio track. When they are all used together, these modality signals are the most powerful features for an analysis of video content to convey meanings and semantic information, as compared to any single modality. In a sport video, audiovisual joint analysis is effective to detect key episodes, such as a goal event scene, touchdown sequence [152], and racket hits in a tennis game [153]. In television broadcasting, multimodality signals can be used to

# Templates	Off-line processing		On-line query execution
	Template generation	AVI indexing	
R=500	1 minutes 6 sec.	1.30 sec.	0.28 sec.
R=1000	2 minutes 2 sec.	2.31 sec.	0.38 sec.
R=1500	5 minutes 7 sec.	4.50 sec.	0.58 sec.
R=2000	6 minutes 45 sec.	5.95 sec.	0.67 sec.

Table 7.4. Processing time used in the online and offline process in iARM system. AVI indexing time was measured from a single video clip of size 0.27 seconds. The off-line process was implemented on a Pentium IV 2.4GHz PC using Window XP system and MATLAB implementation. The on-line process used a Pentium IV 2.0 GHz PC Server system and JAVA implementation. The algorithms associated with the offline process assume that video frames have been described by color histogram feature vectors.

classify video into certain types of scenes such as a dialogue, story, and action [154]; weather forecasts, and commercials [155]. The multimodality also plays an important role in combining content and context features for multimedia understanding. This understanding includes multimedia object detection and organization [156, 157], which aims at supporting a better interface of high-level-concept queries.

Upon the domain-specific application, event and object detection algorithm using audiovisual information can be integrated to facilitate video browsing, filtering, and retrieval [152]- [157]. These high-level techniques are, however, heavily dependent on specific attributes, such as events and objects, which makes the extension for characterization of other unspecified contents very difficult. In practical applications, we need a generic framework that can be effectively applied to the retrieval of various types of videos such as movies, music videos, news, and commercials.

In this section, a work on joining visual and audio descriptors is presented for effective video content indexing and retrieval. Unlike the previous works discussed, the current method retains its *scalability* and *flexibility* properties. Firstly, the scalability demonstrates that the current technique can be applied to a wide range of applications. As compared to the techniques in [152–154], the current method does not restrict itself neither to specific video domains nor pre-defined event and scene classes. Secondly, this technique is flexible in that it can be applied to a longer portion of video clip, beyond the shot or key frames. It is shown in Chapter 6 that the AVI technique is flexible for application to a longer portion of video clip, beyond the shot or key frames.

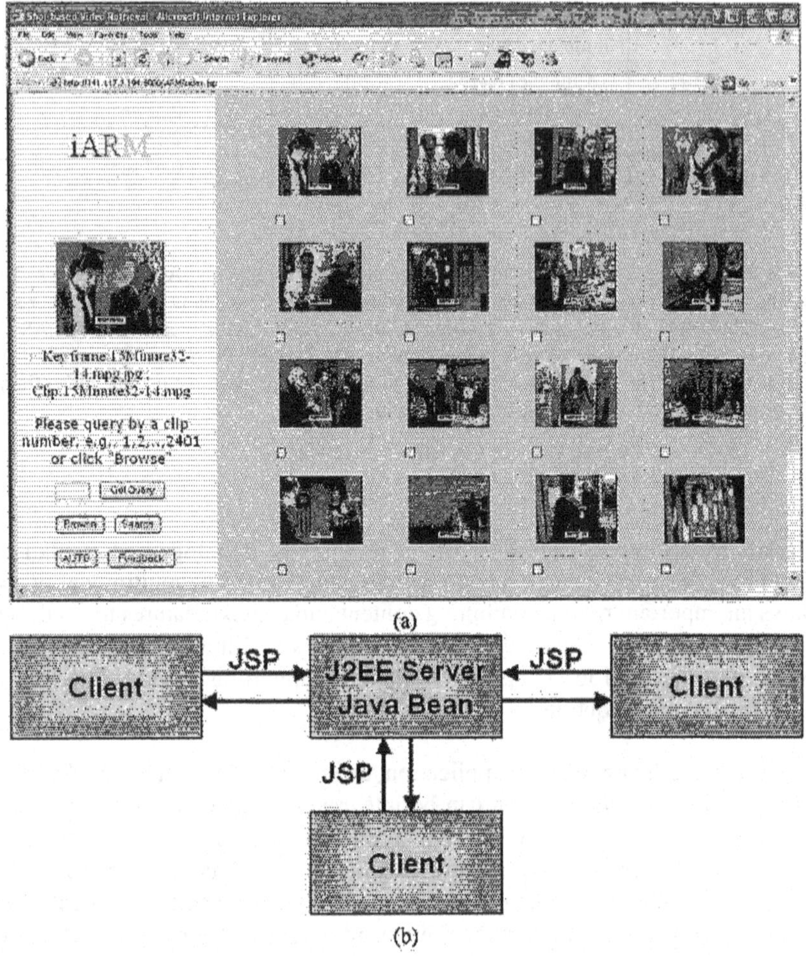

Figure 7.1. (a) iARM's GUI using two query interfaces: a video ID search and a query-by-example method; (b) implementation software diagram.

As well as the visual domain, the current method for the characterization of audio content is flexible enough to apply to an audio clip that contains several applications of music, speech, or noise. A statistical time-frequency analysis method is presented for the characterization of audio clip, where the method is independent to the pre-defined audio segments. The audio is regarded as a non-stationary signal, in which the signal characteristic can be changed dramatically within a given portion of audio clip. Unlike the audio classification techniques proposed by Wold [158] and Saunders [159], which emphasize a few pre-defined audio classes such as speech and music, the current method can attain features that address a high degree of flexibility in characterizing the comprehensive characteristics of the audio content.

The main objective is to demonstrate the analysis techniques of visual and audio contents, and how these visual and audio features can be jointed in an effective manner for video indexing and retrieval. It will shown that the new audiovisual analysis technique can be generalized and effectively applied for retrieval of various video types, including movie, music video, and commercial from a database of 6,000 video clips.

Feature Extraction from Embedded Audio

The users of the video data are often interested in certain action sequences that are easier to identify in the audio domain. While visual information may not yield useful indexes in this scenario, the audio information often reflects what is happening in the scenes. The existing techniques to classify audio or the embedded audio may not be very useful for application to video retrieval. In this type of application, music, speech, noise and crowd voice may be found together in the same video clip. Therefore, it is desirable to have features that represent the global similarity of the audio content. A statistical approach based on wavelet transformation and Laplacian mixture models has been adopted here to analyze the audio data and extract the features for video indexing.

Wavelet coefficient distributions are very peaky in nature due to their energy packing property (see Figure 7.2). This type of peaky distributions is non-Gaussian in nature. We can model any arbitrary shaped distribution using finite mixture model [10]. Here a mixture of only two Laplacians is used for modeling the shape of the distribution. The model parameters of this mixture are used as features for indexing the video clips based on the audio information only.

Feature Extraction Algorithm

The current method for audio content indexing does not depend on the segmentation method. The video may be firstly segmented into clips using any existing algorithm. Audio is then separated from the video clips and the signals are re-sampled to a uniform sampling rate. Each audio segment is decomposed using a one-dimensional DWT. One-dimensional DWT decomposes the audio signals into 2 sub-bands at each wavelet scale; a low frequency subband and a high frequency subband. The wavelet decomposition scheme matches

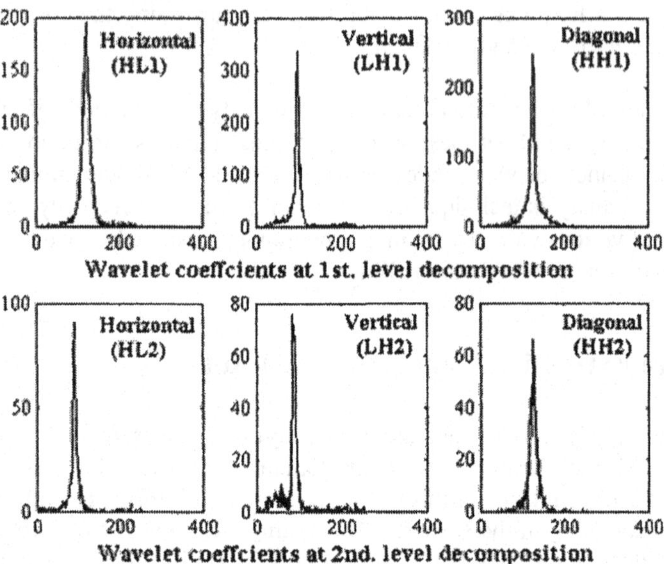

Figure 7.2. Histograms of wavelet coefficients obtained from a texture image, using 2-level decomposition.

the models of sound octave-division for perceptual scales. Wavelet transform also provides a multi-scale representation of sound information, so that we can build indexing structure based on this scale property. Moreover, audio signals are non-stationary signals whose frequency contents evolve with time. Wavelet transform provides both frequency and time information simultaneously. These properties of wavelet transform for sound signal decomposition are the foundation of the current audio based video indexing and retrieval system.

An increase in the level of wavelet decomposition increases the number of features extracted for indexing. This improves the retrieval performance at the expense of more computational overhead. We now model the distribution of the wavelet coefficients in the high frequency sub-bands using a mixture of two Laplacians centered at 0. The parameters of this mixture model are used as features for indexing the texture images. It has been observed that the resulting features possess high discriminatory power for texture classification. Because of the low dimensionality of the resulting feature vector, the retrieval stage consumes less time enhancing the user experience while interacting with the system.

The model can be represented as:

$$p(w_i) = \alpha_1 p_1(w_i|b_1) + \alpha_2 p_2(w_i|b_2), \quad \alpha_1 + \alpha_2 = 1 \tag{7.1}$$

where α_1 and α_2 are the mixing probabilities of the two components p_1 and p_2; w_i are the wavelet coefficients; b_1 and b_2 are the parameters of the Laplacian distribution p_1 and p_2, respectively. The Laplacian component in Eg.(7.1) is defined as:

$$p_1(w_i|b_1) = \frac{1}{2b_1}\exp(-|w_i|/b_1) \tag{7.2}$$

The shape of the Laplacian distribution is determined by the single parameter b. The EM algorithm [160] is applied to estimate the parameters of the model.

E-Step: For the *n-th* iterative cycle, the E-step computes two probabilities for each wavelet coefficient:

$$p_{1i} = \frac{\alpha_1(n)p(w_i|b_1(n))}{\alpha_1(n)p(w_i|b_1(n)) + \alpha_2(n)p(w_i|b_2(n))}, \tag{7.3}$$

$$p_{2i} = \frac{\alpha_2(n)p(w_i|b_2(n))}{\alpha_1(n)p(w_i|b_1(n)) + \alpha_2(n)p(w_i|b_2(n))} \tag{7.4}$$

M-Step: In the M-step, the parameters $[b_1, b_2]$ and *a priori* probabilities $[\alpha_1, \alpha_2]$ are updated.

$$\alpha_1(n+1) = \frac{1}{K}\sum_{i=1}^{K} p_{1i}(n), \quad \alpha_2(n+1) = \frac{1}{K}\sum_{i=1}^{K} p_{2i}(n), \tag{7.5}$$

$$b_1(n+1) = \frac{\sum_{i=1}^{K}|w_i|p_{1i}(n)}{K\alpha_1(n+1)}, \quad b_2(n+1) = \frac{\sum_{i=1}^{K}|w_i|p_{2i}(n)}{K\alpha_2(n+1)} \tag{7.6}$$

The following components form the feature vector used for indexing the video clips:

- Mean and standard deviation of the wavelet coefficients in the low frequency subband.
- Model parameters $[\alpha_1, b_1, b_2]$ calculated for each of the high frequency subband.

For N-level decompositions, the above algorithm gives $(N \times 3) + 2$ feature components for characterized audio content in the input video clip.

Audio-visual Information Fusion

In developing audio-visual information recognition systems, there are two techniques suggested for audio-visual fusion: feature fusion and decision fusion [161]. The first approach is a simple audio-visual feature concatenation, giving rise to a single data representation before the pattern matching stage. The second approach applies separate pattern matching algorithms to audio and video data and then merges the estimated likelihoods of the single-modality matching decisions. In the current application, a decision fusion scheme is chosen due to the following two reasons. First, visual feature vectors in the current work have physical structure different from audio features in both feature dimensions and weighting scheme. Second, based on previous studies in [162] with respect to human perception, audio and visual processing is likely to be carried out independently in different modalities and combined at a very late stage. The second suggests that audio and visual information are orthogonal to some degree. Audio contains information that is often not available in the visual signal [163]; thus, it may be not appropriate to concatenate audio and visual into a single representation.

This work introduces a decision fusion scheme to obtain a similarity score given as:

$$D_t^{(AV)} = Rank_t^{(A)} Rank_t^{(V)} \quad t \in \{1, ..., K\} \tag{7.7}$$

where $Rank_t^{(A)}$ is the ranking of the t-th video by means of the Euclidean distance in the audio-feature space; $Rank_t^{(V)}$ is the ranking of the t-th video by means of the Cosine measure in the visual-feature space; and K is the total number of video files in the database. The distances $D_t^{(AV)} t = 1, ..., K$ are sorted in decreasing order in order to obtain retrieval result.

In previous discussion it has been shown that the principle of the AVI visual indexing structure is based on the term-vector model as in text retrieval. This model allows a query expansion mechanism in a form of relevance feedback to increase retrieval accuracy. More specifically, a given query vector is modified by adjusting positive and negative weights extracted from feedback samples. This relevance feedback technique is now utilized to modify query vectors in order to obtain ranking discussed in Eq. 7.7. After performing a first search and receiving feedback from users, the query vectors are modified by:

$$\mathbf{v}_{New}^{(V)} = \alpha \mathbf{v}_{old}^{(V)} + \beta \bar{\mathbf{x}}_{Post}^{(V)} - \varepsilon \bar{\mathbf{x}}_{Neg}^{(V)} \tag{7.8}$$

$$\mathbf{v}_{New}^{(A)} = \alpha \mathbf{v}_{old}^{(A)} + \beta \bar{\mathbf{x}}_{Post}^{(A)} - \varepsilon \bar{\mathbf{x}}_{Neg}^{(A)} \tag{7.9}$$

where: $\mathbf{v}_{New}$ is a new modified query; $\mathbf{v}_{Old}$ is the query before having modification, $\bar{\mathbf{x}}_{Post}$ is the centriod of positive samples; $\bar{\mathbf{x}}_{Neg}$ is the centriod of negative samples; $(\alpha, \beta, \varepsilon)$ are positive constant values; and the superscriptions (V) and (A) represent visual and audio features, respectively.

In a search process, the system utilizes the user submitted query to obtain ranking from audiovisual fusion formula as in Eq. (7.7), followed by relevance feedback from user. The positive and negative samples provided by the user are then used for query modification in visual and audio feature space as discussed in Eqs. (7.8) and (7.9). The system again utilize Eq.(7.7) to obtain a new video ranking from the newly modified queries.

Retrieval of Movie Clips Using Audiovisual Fusion

Experiments were conducted on a database containing 3 Hollywood movies, 15 music videos, and 6 commercial videos. These videos were segmented into 6000 clips, each of which contained one to three shots at approximately 30 seconds in length. For each clip, a color histogram was extracted as a visual feature vector from each frame. A number of histogram vectors randomly selected from 6000 video clips were used for training by vector quantization process. The number of templates used in the AVI algorithm was set to 2000, and the templates were used by a mapping process for obtaining visual feature vector. By a Matlab implementation, the system took approximately 6 seconds for indexing a single video clip. In obtaining audio feature vectors, a wavelet transform with 9-level decompositions was applied to audio signal from each video clip. The wavelet coefficients in each subband were then characterized by LMM, and the resulting model parameters were used to obtain a 29-dimensional feature vector. In this way, as feature components represent different physical quantities and have different dynamic ranges, Gaussian normalization technique discussed in [5] was employed to convert the component feature to [-1, 1].

Five query concepts were utilized to test the system. These concepts include ‘fighting’, ‘ship crashing’, ‘love scene’, ‘music video’, and ‘dancing party’. For each concept, five video clips were used as queries and the retrieval performance was measured by precision rate from the top 16 best matches. The performances were measured at two stages: after a simple search via similarity ranking, and after one cycle of relevance feedback. Table 7.5 presents retrieval results, where precisions are averaged within query concepts, as well as within the overall queries. The results clearly display the benefits of audiovisual fusion technique in retrieving of videos according to the defined concepts; the average precision rate obtained was at 94.0% . Ether audio or visual plays an important

Query Concepts	**Visual feature**	**Audio feature**	**Audio-visual cue**	
			Before feedback	**After feedback**
'Love Scene'	77.5	17.5	80.0	83.8
'Music Video'	73.8	83.8	96.3	100
'Fighting'	57.5	83.8	96.3	98.8
'Ship Crashing'	75.0	65.0	97.5	100
'Dancing Party'	92.5	78.75	100	100
Average	**75.3**	**65.8**	**94.0**	**96.5**

Table 7.5. Average precision rate (%) obtained by using audio, visual, and audiovisual descriptors, measured from five high-level query categories.

role in characterizing the specific concept better than the other. Particularly, audio feature gave better precision rates for retrieval of video clips from the following two concepts: 'music video' and 'fighting'. The characteristic of sound in 'fighting' concept, such as gun shooting, crowd, and hours driving, make the audio feature more useful than visual feature in retrieving this concept. Relevance feedback can be used to improve retrieval result after the first search. In general, it would be a time consuming process if the feedback run across many system cycles. However, the structure of visual feature makes it is suitable for query expansion process (through Eqs.(7.8)-(7.9)), within only a few cycles of relevance feedback. The results after a single feedback from this system was at 96.8% (i.e., 15 relevant videos were retrieved from the top 16 best matches). This result also shows that the audiovisual cue is highly suitable for integrating with relevance feedback in characterizing concepts for video retrieval.

4. Chapter Summary

This chapter presents applications of adaptive video indexing (AVI) technique for movie retrieval. AVI is demonstrated by an online multimedia search engine for retrieval using the query-by-video clip method, and compared to other popular video indexing and retrieval techniques. The search engine implements AVI in both user-controlled and self-organizing relevance feedback processes, which provides very high accuracy retrieval results as well as quick query response time. The chapter also presents an integration of AVI with an audio-visual fusion model that can support high-level concept queries. This was successfully applied for retrieval of movie clips from a large digital collection.

References

[1] M. Flickner, H. Sawhney, W. Niblack, "Query by image and video content: The QBIC system," *IEEE Computer*, vol. 28, no. 9, pp. 23-31, Sept. 1995.

[2] J. R. Smith and S.-F. Chang, "VisualSeek: A fully automated content-based image query system," *Proc. ACM Multimedia Conf.*, MA, USA, pp. 87-98, 1996.

[3] T. Gevers and W. M. Smeulders, "PicToSeek: combining color and shape invariant features for image retrieval," *IEEE Trans. on Image Processing*, vol. 9, no. 1, pp. 102-119, 2000.

[4] A. Celentano and E. D. Sciasicio, "Feature interation and relevance feedback analysis in image similarity evaluation", *Journal of Electronic Imaging*, vol. 7, no. 2, pp. 308-317, 1998.

[5] Y. Rui, T. S. Hang, S. Mehrotra, and M. Ortega, "A relevance feedback architecture for content-based multimedia information retrieval systems," *Proc. IEEE Workshop on Content-based Access of Image and Video Libraries*, pp. 82-89, 1997.

[6] J. Peng, B. Bhanu, S. Qing, "Probabilistic feature relevance learning for content-based image retrieval," *Computer Vision and Image Understanding*, vol. 75, nos. 1/2, pp. 150-164, 1999.

[7] G. Ciocca, R. Schettini, "Using a relevance feedback mechanism to improve content-based image retrieval," in P. Huijsmans and W. M. Smeulders editor, *Visual Information and Information Systems*, Springer, pp. 105-114, 1999.

[8] Y. Rui, T. S. Huang, M. Ortega, and S. Mehrotra, "Relevance feedback: A power tool for interactive content-based image retrieval," *IEEE Trans. Circuits Syst. Video Technol.*, vol. 8, no. 5, pp. 644-655, 1998.

[9] Y. Rui, T.S. Huang, and S. Mehrotra, "Content-based image retrieval with relevance feedback in MARS," *Proc. IEEE Int. Conf. on Image Processing*, Washington D.C., USA, pp. 815-818, 1997.

[10] E. Di Sciascio, M. Mongiello, "DrawSearch: a tool for interactive content-based image retrieval over the net," *Proc. of SPIE*, vol. 3656, pp. 561-572, 1999.

[11] H. Müller, W. Müller, S. Marchand-Maillet, T. Pun, and D. M. Squire, "Strategies for positive and negative relevance feedback in image retrieval," *Int. Conf. on Pattern Recognition*, Barcelona, Spain, vol.1, pp. 1043-1046, Sept., 2000

[12] S. Haykin, *Neural Networks: A Comprehensive Foundation*, Macmillan, New York, 1994.

[13] J. Moody and C. J. Darken, "Fast learning in networks of locally-tuned processing units," *Neural Computation*, vol. 1, no. 2, pp. 281-294, 1989.

[14] Y. Rui and Thomas S. Huang, "Optimizing Learning In Image Retrieval," in *Proc. of IEEE Int. Conf. on Computer Vision and Pattern Recognition*, Hilton Head, SC, vol.1, pp. 236-243, June 2000.

[15] P. Muneesawang and L. Guan, "Multiresolution-histogram indexing for wavelet-compressed images and relevant feedback learning for image retrieval," *Proc. IEEE Int. Conf. on Image Processing*, Vancouver, Canada, vol. 2, pp.526-529, 2000.

[16] S. Sclaroff, L. Taycher, and M. L. Cascia, "ImageRover: A content-based image browser for the world wide web," *Proc. IEEE Workshop on Content-Based Access of Image and Video Libraries*, pp. 2-9, 1997.

[17] R. L. De Valois and K. K. De Valois, *Spatial Vision*, Oxford Science Publications, New York, 1988.

[18] P. Muneesawang and L. Guan, "A nonlinear RBF model for interactive content-based image retrieval," *The First IEEE Pacific-Rim Conf. on Multimedia*, Sydney, Australia, pp. 188-191, 2000.

[19] A. N. Tikhonov, "On solving incorrectly posed problems and method of regularization," In S. Haykin editor, *Neural networks: a Comprehensive Foundation*, Prentice Hall, 1999.

[20] T. Poggio and F. Girosi, "Networks for approximation and learning," *Proceeding of the IEEE*, vol. 78, pp. 1481-1497, 1990.

[21] J. H. Friedman, "Flexible metric nearest neighbor classification." *Technical Report*, Department of Statistics, Standford University, 1994.

[22] Corel Gallery Magic 65000, "www.corel.com", 1999.

[23] A. Gersho and R.M. Gray, *Vector Quantization and Signal Compression*, Norwell, MA:Kluwer, 1992.

[24] B. S. Manjunath and W. Y. Ma, "Texture features for browsing and retrieval of image data," *IEEE Trans. of Pattern Analysis and Machine Intelligence*, vol. 18, no. 8, pp. 837-842, 1996.

[25] G. Salton and M. J. McGill, *Intoduction to Modern Information Retrieval*, McGraw-Hill Book Company, NY, 1983.

[26] S. J. Cunningham, G. Holmes, J. Littin, R. Beale, and I.H. Witten, "Applying connectionist models to information retrieval," in S.-I. Amari and N. Kasabov, *Blain-Link Computing and Intelligent Information Systems*, Springer, 1998.

[27] J. J. Rocchio, "Relevance feedback in information retrieval," In G. Salton, editor, *The SMART Retrieval System ¾ Experiments in Automatic Document Processing*, Prentice Hall Inc., Englewood Cliffs, NJ, 1971.

[28] N. B. Karayiannis, P.-I. Pai, and N. Zervos, "Image compression based on fuzzy algorithms for learning vector quantization and wavelet image decomposition," *IEEE Trans. on Image processing*, vol. 7, no. 8, pp. 1223-1230, 1998.

[29] B. S. Manjunath, P. Wu, S. Newsam, and H. D. Shin, "A texture descriptor for browsing and similarity retrieval," *Signal Processing Image Communication*, vol. 16, Issue 1-2, pp. 33-43, Sept 2000.

[30] P. Wu, B.S.Manjunath, S.D. Newsam and H.D.Shin, "A texture descriptor for image retrieval and browsing", *Computer Vision and Pattern Recognition Workshop*, Fort Collins, CO, USA, June 1999.

[31] M. Safar and C. Shahabi and X. Sun, "Image Retrieval by Shape: A Comparative Study", *IEEE International Conf. on Multimedia and Expo (I)*, New York, USA, pp. 141-144, 2000.

[32] Y. Chen, X. Zhou, and T. Huang, "One-class SVM for learning in image retrieval," *Proc. of IEEE Int. Conf. on Image Processing*, Thessaloniki, Greece, vol.1, pp. 34-37, October 2001.

[33] P. Muneesawang and L. Guan, "A neural network approach for learning image similarity in adaptive CBIR," *Proc. of IEEE Workshop on Multimedia Signal Processing*, Canes, France, pp.257-262, October 2001.

[34] J. Randall, L. Guan, X. Zhang, and W. Li, "Investigations of the Self-Organizing Tree Map," *Proc. of Int. Conf. on Neural Information Processing*, vol 2, pp.724-728, November 1999.

[35] H. Kong and L. Guan, "Self-organizing tree map for eliminating impulse noise with random intensity distributions", *Journal of Electronic Imaging*, vol.7, no.1, pp.36-44, Jan. 1998.

[36] H. Kong, "Self-organizing tree map and its applications in digital image processing", *PhD Thesis*, University of Sydney, 1998.

[37] MPEG Document 2929, Description of Color/Texture core experiments, Melbourne, Australia, October 1999.

[38] J.A. Lay and L. Guan, "Image retrieval based on energy histogram of the low frequency DCT coefficients," *Proc. of IEEE Int. Conf. on Accoustic Speech and Signal Processing*, Phoenix, USA, pp. 3009-3012, 1999.

[39] Media Graphics International, "Photo Gallery 5,000," vol.1 CD-ROM, www.media-graphics.net.

[40] F. Arman, R. Depommier, A. Hsu, and M.-Y. Chiu, "Content-based browsing of video sequences," *Proc. ACM Multimedia*, San Francisco, USA, pp.97-103, 1994.

[41] H. J. Zhang, S. W. Smoliar, and J. H. Wu, "Content-based video browsing tools," *Proc. SPIE Storage and Retrieval for Image and Video Databases*, Philadelphia, USA, pp. 389-398, 1995.

[42] W.A. Harry and C.A. Marios, "Modeling content for semantic-level querying of multimedia," *Multimedia Tools and Application*, Kluwer Academic Publishers, Vol. 15, No. 1, pp. 5-37, Sept 2001.

[43] R. Wilkinson and P.Hingston, "Using the cosine measure in a neural network for document retrieval," *Proc. of the ACM SIGIR Conf. on Research and Development in Information Retrieval*, pp. 202-210, Chicago, USA, Oct 1991.

[44] G. Salton, E.A. Fox, and E. Voorheers, "Advanced feedback methods in information retrieval," *Journal of the American Society for Information science*, Vol. 36, No. 3, pp. 200-210, 1985.

[45] M. Naphades, R. R. Wang, T. Huang, "Audio-visual query and retrieval: a system that uses dynamic programming and relevance feedback" *Journal of Electronic Imaging*, pp.861-870 Oct 2001.

[46] T. Kohonen, *Self-Organizing MAPS.* 2nd ed., Springer-Verlag, Berlin, 1997.

[47] U. Gargi, R. Kasturi, and S.S. Strayer, "Performance characterization of video-shot-change detection methods," *IEEE Trans. on Circuits and Systems for Video Technology*, Vol. 10, No. 1, pp. 1-13, Jan 2000.

[48] Informedia Digital Video Library Project at Carnegie Mellon University, http://www.informedia.cs.cmu.edu, Sept 2001.

[49] S.-F. Chang and H. Sundaram, "Structural and semantic analysis of video," *IEEE Int. Conf, on Multimedia and Expo*, New York, USA, vol.2, pp. 687-690, July 2000.

[50] T. Wang, Y. Rui, and S.-M. Hu, "Optimal adaptive learning for image retrieval," *IEEE Computer Society Conference on Computer Vision and Pattern Recognition*, Hawaii, USA, vol.1, pp. 1140-1147, 2001.

[51] X. S. Zhou and T. S. Huang, "Exploring the nature and variants of relevance feedback," *IEEE Workshop on Content-Based Access of Image and Video Libraries*, Hawaii, USA, pp.94-101, 2001.

[52] J. Yoon and N. Jayant, "Relevance feedback for semantics based image retrieval," *IEEE Int. Conf. on Image Processing*, Thessaloniki, Greece, vol.1, pp. 42-45, 2001.

[53] N. Doulamis and A. Doulamis, "A recursive optimal relevance feedback scheme for content based image retrieval," *IEEE Int. Conf. on Image Processing*, Thessaloniki, Greece, vol.2, pp. 741-744, 2001.

[54] C. de Mauro, M. Gori, and M. Maggini, "APEX an adaptive visual information retrieval system," *IEEE Int. conf. on Document Analysis and Recognition*, Seattle, USA, pp. 898-902, 2001.

[55] N. Vasconcelos and A. Lippman, "Bayesian relevance feedback for content-based image retrieval," *IEEE Workshop on Content-Based Access of Image and Video Libraries*, Hilton Head, SC, pp. 63-67, 2000.

[56] Y. Wu, Q. Tian, and T. S. Huang, "Integrating unlabeled images for image retrieval based on relevance feedback," *IEEE Int. conf. on Pattern Recognition*, Barcelona, Spain, vol.1, pp. 21-24, 2000.

[57] Y. Wu, Q. Tian, T. S. Huang, "Discriminant-EM algorithm with application to image retrieval," *Proc. IEEE Computer Vision and Pattern Recognition*, vol.1, pp. 222-227, 2000.

Multimedia and Expo, 2000. ICME 2000. 2000 IEEE International Conference on , Volume: 1 , 2000 Page(s): 331 -334 vol.1

[58] N. D. Doulamis, A. D. Doulamis, and S. D. Kollias, "Non-linear relevance feedback improving the performance of content based retrieval systems," *IEEE Int. Conf. on Multimedia and Expo*, vol. 1, pp. 331-334, 2000.

[59] W. El-Naqa, M. N. Wernick, Y. Yand, and N. P. Galatsanos, "Image retrieval based on similarity leaning," *Prod. IEEE Int. Conf. on Image Processing*, Vanouver, Canada, vol.3, pp. 722-725, 2000.

[60] M. E. J. Wood, N. W. Campbell, and B. T. Thomas, "Iterative refinement by relevance feedback," *ACM Multimedia*, Bristol, UK, pp. 13-20, 1998.

[61] J. Laaksonen, M. Koskela, and E. Oja, "PicSom-self-organizing image retrieval with MPEG-7 content descriptions," *IEEE Trans. on Neural Network*, vol. 13, no. 4, pp. 841-853, July 2002.

[62] Y. Chen, X. S. Zhou, and T. S. Huang, "One-class SVM for learning in image retrieval," *IEEE Int. Conf. on Image Processing*, Thessaloniki, Greece, vol.1, pp. 34-37, October 2001.

[63] L. Zhang, F. Lin, B. Zhang, "Support vector machine learning for image retrieval," *IEEE Int. Conf. on Image Processing*, Thessaloniki, Greece, vol.2, pp. 721-724, October 2001.

[64] J. Peng, "A multi-class relevance feedback approach to image retrieval," *IEEE Int. Conf. on Image Processing*, Thessaloniki, Greece, vol.1, pp. 46-49, October 2001.

[65] P. Wu and B. S. Manjunath, "Adaptive nearest neighbor search for relevance feedback in large image databases," *Proc. ACM International Multimedia Conference,* Ottawa, Canada, Oct. 2001.

[66] S. Aksoy, R. M. Haralick, F. A. Cheikh, M. Gabbouj, "A weighted distance approach to relevance feedback," *IEEE Int. Conf. on Pattern Recognition*, Barcelona, Spain, vol.4, pp. 812-815, 2000.

[67] D. Bhanu, J. Peng, and S. Qing, "Learning feature relevance and similarity metrics in image database," *IEEE Workshop on Content-Based Access of Image and Video Libraries*, CA, USA, pp. 14-18, 1998.

[68] L. Wu, C. Faloutsos, K. Sycara, and T. R. Payne, "FALCON: Feedback adaptive loop for content-based retrieval," *Proc. VLDB conf.*, Cairo, Egypt, 2000.

[69] Y.-S. Choi, D. Kim, and R. Krishnapuram, "Relevance feedback for content-based image retrieval using the choquet integral," *IEEE Int. Conf. on Multimedia and Expo*, New York USA, vol.2, pp. 1207-1210, 2000.

[70] S. Sclaroff, L. Taycher, and M. La Cascia, "ImageRover: A content-based image browser for the World Wide Web," *Proc. IEEE Workshop on Content-based Access of Image and Video Libraries*, pp. 2-9, June 1997.

[71] T. V. Ashwin, N. Jain, S. Ghosal, "Improving image retrieval performance with negative relevance feedback," *IEEE Int. Conf. on Acoustics, Speech, and Signal Processing*, Salt Lake City, USA, vol.3, pp. 1637-1640, 2001.

[72] G. Giacinto, F. Roli, and G. Fumera, "Comparison and combination of adaptive query shifting and feature relevance learning for content-based image retrieval," *IEEE Int. Conf. on Image Analysis and Processing*, Palermo, Italy, pp. 422-427, 2001.

[73] S. J. Yoon, D. K. Park, S-J. Park, and C. S. Won, "Image retrieval using a novel relevance feedback for edge histogram descriptor of MPEG-7," *IEEE Int. conf. on Consumer Electronics*, pp. 354-355, 2001.

[74] D. R. Heisterkamp, J. Peng, and H. K. Dai, "Feature relevance learning with query shifting for content-based image retrieval," *IEEE Int. Conf. on Pattern Recognition*, Barcelona, Spain, vol,4, pp. 250-253, 2000.

[75] B. Patrice and H. Konik, "Texture similarity queries and relevance feedback for image retrieval," *IEEE Int. Conf. on Pattern Recognition*, Barcelona, Spain, vol,4, pp. 55-58, 2000.

[76] Y. Ishikawa and R. Subramanya, "MindReader: Query database through multiple examples," *Proc. of Int. Conf. on Very Large Data Bases*, New York, USA, 1998.

[77] I. J. Cox, M. L. Miller, T. P. Minka, T.V. Parathomas, and P. N. Yianilos, "The Bayesian image retrieval system, PicHunter: Theory, implementation, and psychophysical," *IEEE Trans. on Image Processing*, vol. 9, no. 1, pp. 102-119, Jan. 2000.

[78] Y. Zhuang, X. Liu, Y. Pan, "Apply semantic template to support content-based image retrieval," *Proc. SPIE Storage and Retrieval for Multimedia Database*, USA, pp. 442-449, Jan. 2000.

[79] S. Chen, C. F. N. Cowan, and P. M. Grant, "Orthogonal least squares learning algorithm for radial basis function networks," *IEEE Trans. on Neural Networks*, vol. 2, no. 2, March 1991.

[80] H. Müller, W. Müller, S. Marchand-Maillet, T. Pun, and D. McG. Squire, "Automated benchmarking in content-based image retrieval," *Proc. of the IEEE International Conf. on Multimedia and Expo* , Tokyo, Japan, October 22-25 2001.

[81] T. Poggio and F. Girosi, "A theory of networks for approximation and leaning," *Technical Report* A.I. Memo No. 1140, Massachusetts Institute of Technology, 1989.

[82] D. S. Broomhead and D. Lowe, "Multivariable functional interpolation and adaptive networks," *Complex Syst.* , vol. 2, pp. 321-355, 1988.

[83] S. Chen, P.M. Grant and C.F.N. Cowan, "Orthogonal least squares algorithm for training multi-output radial basis function networks," *IEE Proc. Part F*, vol.139, no.6, pp.378-384, 1992.

[84] J. Park and I. W. Sandberg, "Universal approximation using radial-basis function networks," *Neural Computation*, vol.3, pp. 246-257, 1991.

[85] S. Haykin, "Neural networks: a comprehensive foundataion," Prentice Hall, Upper Saddle River, New Jersey, 1999.

[86] MathWorks, Inc. "Neural network toolbox user's guide," Version 6 (R12), 2001.

[87] R. Murray-Smith and T. A. Johansen, "Local learning in local model networks," *Multiple Model Approaches to Modelling and Control*, Taylor and Francis, pp 185-210, 1997.

[88] B. S. Manjunath, J.R. Ohm, V. V. Vasudevan, and A. Yamada, "Color and Texture Descriptors," *IEEE Trans. on Circuit and Systems for Video Technology*, vol. 11, no. 6, pp. 703-715, June 2001.

[89] W. Ma and B. Manjunath, "EdgeFlow: a technique for boundary detection and image segmentation," *IEEE Trans. on Image Processing*, vol. 9(8), pp. 1375-88, August 2000.

[90] F. Jing, B. Zhang, F. Lin, W.-Y. Ma, H.-J. Zhang, "A novel region-based image retrieval method using relevance feedback," *3rd Intl Workshop on Multimedia Information Retrieval*, Ottawa, Canada , October 2001

[91] J. Li, J. Z. Wang, and G. Wiederhold, "IRM: integrated region matching for image retrieval," *ACM Multimedia*, Los Angeles, USA, pp. 147-156, 2000.

[92] X. S. Zhou and T. Huang, "Unifying keywords and visual contents in image retrieval," *IEEE Multimedia*, vol. 9 no. 2, pp. 23-33, April-June 2002.

[93] M. La Cascia, S. Sethi, and S. Sclaroff, " Combining textual and visual cues for content-based image," *IEEE Workshop on Content-based Access of Image and Video Libraries*, Los Angeles, USA, pp. 24-28, June 1998.

[94] J. Randall, L. Guan, X. Zhang, and W. Li, "Hierarchical cluster model for perceptual image processing," *Proc. IEEE Int. Conf. Acoustics, Speech, and Signal Processing*, vol. 1pp. 1041 - 1044, May 13-17, 2002

[95] M. K. Mandal, S. Panchanathan, and T. Aboulnasr, "Image Indexing Using Translation and Scale-Invariant Moments and Wavelets," *Storage and Retrieval for Image and Video Databases (SPIE)*, pp. 380-389, 1997.

[96] C. Faloutsos, and K. Lin, "Fastmap: A fast algorithm for indexing, data mining and visualization of traditional and multimedia," *Proc. of ACM Special Interest Group on Management of Data* pp. 163-174, 1995.

[97] H.-S. Wong, M. Wu, R. A. Joyce, L. Guan, and S.-Y. Kung, "A neural network approach for predicting network resource requirement in video transmission systems," *The First IEEE Pacific-Rim Conf. on Multimedia*, Sydney, Australia, pp. 116-119, Dec. 2000.

[98] P. Salembier and J. R. Smith, "MPEG-7 multimedia descriptor schemes," *IEEE Trans. on Circuits and Systems for Video Technology*, vol. 11, no. 6, pp. 748-759, June 2001.

[99] A. Jain and G. Healey, "A multiscale representation including opponent color features for texture recognition," *IEEE Trans. on Image Processing*, vol. 7, pp. 124-128, Jan. 1998.

[100] P. Bocheck and S. F. Cheng, "Content-based VBR traffic modeling and its application to dynamic network resource allocation," *Teck. Rep.* 48c-98-20, Columbia University, 1998.

[101] ISO/IEC, ISO/IEC 14496-2:1999: Information technology—coding of audio-visual objects—Part 1: visual, Dec. 1999.

[102] ISO/IEC JTC 1/SC 29/WG 1, ISO/IEC FDIS 15444-1: information technology—JPEG 2000 image coding system: core coding system [WG 1 N 1890], Sept. 2000.

[103] A. Said, W. A. Pearlman, "A new and efficient image codec based on set partitioning in hierarchical trees," *IEEE Trans. Circuits Systems Video Technol.*, vol. 6, no. 3, pp. 243-250, June 1996.

[104] Z. Xiong and T. S. Huang, "Subband-based, memory-efficient JPEG2000 images indexing in compressed-domain," *IEEE Southwest Symposium on Image Analysis and Interpretation*, Santa Fe, USA, April 2002.

[105] C. Lui and M. K. Mandal, "Fast image indexing based on JPEG2000 packet header," *Proc. Int. Workshop on Multimedia Information Retrieval*, Oct. 2001.

[106] M. N. Do and M. Vertterli, "Wavelet-based texture retrieval using generalized Gaussian density and Kullback-Leibler distance," *IEEE Trans. on Image Processing*, vol. 11, no. 2, pp. 146-158, February 2002.

[107] D.-G. Sim, H.-K. Kim and R.-H. Park, "Fast texture desciption and retrieval of DCT-based compressed images," *IEEE Electronic Letters*, vol. 37, no. 1, pp. 18-19, Jan. 2001.

[108] Z. Xiong and T. S. Huang, "Fast, memory-efficient JPEG2000 images indexing in compressed domain," *IEEE Southwest Symposium on Image Analysis and Interpretation*, Santa Fe, USA, April 2002.

[109] J. Bhalod, G. F. Fahmy, and S. Panchanathan, "Region based indesing in the JPEG2000 framework," *Proc. Int. Workshop on Content-based Multimedia Indexing*, Brescia, Italy, Sept. 2001.

[110] M. Antonini, M. Barlaud, F. Mathieu, and I. Daubechies, "Image coding using wavelet transform," *IEEE Trans. on Image Processing*, vol. 1. no. 2, pp. 205-220, April 1992.

[111] T. W. Ryan, L. D. Sanders, H. D. Fisher, and A. E. Iverson, "Image compression by texture modeling in the wavelet domain," *IEEE Trans. on Image Processing*, vol. 5, no. 1, pp. 26-36, Jan. 1996.

[112] D. S. Santas-Cruz, R. Grosbois, and T. Ebrahimi, "JPEG 2000 performance evaluation and assessment," *Signal Processing: Image Communication*, vol. 17, no. 1, 2002.

[113] R. Baeza-Yates and B. Ribeiro-Neto, *Modern Information Retrieval*, ACM Press, New York, 1999.

[114] Y. Ishikawa, R. Subramanya, and C. Faloutsos, "Mindreader: query database through multiple examples," *Prof. of Int. Conf. on Very Large Data Bases* New York, 1998.

[115] *MPEG-7 Visual Experimentation Model (XM)*, Version 10.0, ISO/IEC/JTC1/SC29/WG11, Doc. N4063, March 2001.

[116] MIT Database, ftp://whilechapel.media.mit.edu/pub/VisTex/, 2000.

[117] *Text of ISO/IEC 15 938-3 Multimedia Content Description interface—Part 3: Visual. Final Committee Draft*, ISO/IEC/JTC1/SC29/WG11, Doc. N4062, Mar. 2001.

[118] G. M. Haley and B. S. Manjunath, "Rotation invariant texture classification using a complete space-frequency model," *IEEE Trans. Image Processing*, vol. 8, pp. 255-269, 1999.

[119] Y. Rui, "Efficient indexing, browsing and retrieval of image/video content," PhD Thesis, University of Illinois, 1999

[120] H. Start and J. W. Woods, *Probability, Random Processes, and Estimation Theory for Engineers*, Pretic-Hall, 1986.

[121] M. R. Naphade and T. S. Huang, "A probabilistic framework for semantic video indexing, filtering and retrieval over the Internet," *IEEE Trans. on Multimedia*, vo. 3, pp. 141-151, 2001

[122] R. Fablet, P. Bouthemy, and P. Perez, "Non-parametric motion characterization using causal probabilistic models for video indexing and retrieval," *IEEE Trans. on Image Processing*, vol. 11, no. 4, pp. 393-407, April 2002.

[123] S.-F. Chang and H. Sundaram, "Structural and semantic analysis of video," *Int. Conf. on Multimedia and Expo*, New York, USA, vol.2, pp. 687-690, July 2000.

[124] U. Gargi, R. Kasturi, and S. H. Strayer, "Performance characterization of video-shot-change detection methods," *IEEE Trans. on Circuits and Systems for Video Technology*, vol. 10, no.1, pp. 1-13, Feb. 2000.

[125] R. Wang, M. R., and T. S. Huang, "Video retrieval and relevance feedback in the context of a post-integration model," *IEEE Int. Workshop on Multimedia Signal Processing*, Cannes, France, pp. 33-38, 2001.

[126] N. Haering, R. J. Qian, and M. I. Sezan, "A semantic event-detection approach and its application to detecting hunts in wildlife video," *IEEE Trans. on Circuits and Systems for Video Technology*, vol. 10, no. 6, Sept. pp. 857-868, 2000.

[127] H. Sundaram and S.-F. Chang, "Determining computable scenes in films and their structures using audio-visual memory models," *ACM Multimedia*, Los Angeles, USA, 2000.

[128] X. Tang, X. Gao, J. Liu, and H. Zhang, "A spatial-temporal approach for video caption detection and recognition," *IEEE Trans. on Neural Network*, Vol.13, Issue 4, pp. 961-971, July 2002.

[129] M. A. Smith and T. Kanada, "Video skimming and characterization through the combination of image and language understanding technique," *Proc. IEEE Conf. on Computer Vision and Pattern Recognition*, pp. 775-781, 1997.

[130] S. Basu, M. Naphade, and J. R. Smith, "A statistical modeling approach to content based retrieval," *Proc. of IEEE Int. Conf. on Accoustic Speech and Signal Processing*, vol. IV, pp. 4080-4083, 2002.

[131] E. Sahouria and A. Zakhor, "Content analysis of video using principal components," *IEEE Trans. on Circuits and Systems for Video Technology*, vol. 9, no. 8, pp. 1290-1298, Dec., 1999.

[132] C.-W. Ngo, T.-C. Pong, and H.-J. Zhang, "On clustering and retrieval of video shots," *ACM Multimedia*, Ottawa, Canada, pp. 51-60, 2001.

[133] S.-H. Jeong, J.-H. Choi, and J.-D. Yang, "A concept-based video retrieval model with spatio-temporal fuzzy triples," *IEEE Region 10 Int. Conf. on Electrical and Electronic Technology*, vol.1, pp. 424-429, 2001.

[134] J. Lee and B. W. Dickinson, "Hierarchical video indexing and retrieval for subband-coded video," *IEEE Trans. on Circuits and Systems for Video Technology*, vo. 10, no. 5, pp. 824-829, August 2000.

[135] H. S. Chang, S. Sull, and S. U. Lee, "Efficient video indexing scheme for content-based retrieval," *IEEE Trans. on Circuits and Systems for Video Technology*, vo. 9, no. 8, pp. 1269-1279, Dec. 1999.

[136] A. K. Jain, A. Vailaya, and W. Xiong, "Query by video clip," *Multimedia Systems Journal*, vol.7, Issue 5, pp. 369-384, 1999.

[137] S. Pfeiffer, "Scene determination based on video and audio features," *Multimedia Tools and Application*, vol. 15, pp. 59-81, 2001.

[138] D. Zhong and S.-F. Chang, "An integrated approach for content-based video object segmentation and retrieval," *IEEE Trans. on Circuits and Systems for Video Technology*, vo. 9. no. 8, pp. 1259, pp. 1259-1268, Dec. 1999.

[139] R. Castogno, T. Ebrahimi, and M. Kunt, "Video segmentation based on multimedia features for interactive multimedia applications," *IEEE Trans. on Circuits and Systems for Video Technology*, vo. 8, no. 5, pp. 562-571, Sept. 1998.

[140] H.-Bong Kang, "A hierarchical approach to scene segmentation," *IEEE Workshop on Content-Based Access of Image and Video Libraries*, pp. 65-71, 2001.

[141] X. Gao and X. Tang, "Unsupervised and model-free news video segmentation," *IEEE Workshop on Content-Based Access of Image and Video Libraries*, Hawaii, USA, pp. 58-64, 2001.

[142] A. Hanjalic, R. L. Lagendijk, and J. Biemond, "Automated high-level movie segmentation for advanced video-retrieval systems," *IEEE Trans. on Circuits and Systems for Video Technology*, vo. 9, no. 4, pp. 580-588, June 1999.

[143] H. Zhang, Y. Gong, S. W. Smoliar, "Automatic parsing of news video," *Proc. Int. Conf. on Multimedia Computing and Systems*, Boston, USA, pp. 45-54, 1994.

[144] http://iarm.ee.ryerson.ca, 2002.

[145] C. H. Ng, and K. C. Sia, "Peer Clustering and Firework Query Model," *Proc. of the 11th World Wide Web Conference*, Honolulu, USA, May 2002.

[146] J. Lay, and L. Guan, "SOLO: an MPEG-7 optimum search tool," *IEEE Inter-national Conference on Multimedia and Expo*, Aug 2003.

[147] H. S. Nwana, "Software Agents: An Overview," *Knowledge Engineering Review*, vol. 11, no. 3, pp.1-40, 1996.

[148] W. Y. Ma and B. S. Manjunath, "Edge flow: a framework for boundary detection and image segmentation," *Proc. IEEE International Conference on Computer Vision and Pattern Recognition*, pp. 744-749, Puerto Rico, 1997.

[149] H.-S. Wong and L. Guan, "Characterization for perceptual importance for object-based image segmentation," *Proc. IEEE Int. Conf. on Image Processing*, pp. 54-57, Vancouver, 2000.

[150] D. Mukherjee, Y. Deng and S.K. Mitra, "A region-based video coder using edge flow segmentation and hierarchical affine region matching," *Proc. of SPIE*, vol. 3309, 1998.

[151] P. Muneesawang and L. Guan, "Image retrieval with embedded sub-class information using Gaussian mixture models," *Proc. IEEE Int. Conf. on Multimedia and Expo*, Maryland, USA, pp. 769-772, vol.1, July 2003.

[152] Y.-L. Chang, W. Zeng, I. Kamel, and R. Alonso, "Integrated image and speech analysis for content-based video indexing," *Proc. of IEEE Int. Conf. on Multimedia Computing and Systems*, pp. 306-313, 1996.

[153] R. Dahyot, A. Kokaram, N. Rea, and H. Denman, "Joint audio visual retrieval for tennis broadcasts," *Proc. of IEEE Int. Conf. on Acoustics, Speech, and Signal Processing*, vol. 3, pp. 561-564.

[154] C. Saraceno, "Video content extraction and representation using a joint audio and video processing," *Proc. of IEEE Int. Conf. on Acoustics, Speech, and Signal Processing*, vol. 6, pp. 3033-3036, 1999.

[155] J. Huang, Z. Liu, Y. wang, Y. Chen, and E. K. Wong, "Integration of multimodal features for video scene classification based on HMM," *IEEE Workshop on Multimedia Signal Processing*, pp. 53-58, 1999.

[156] R. S. Jasinschi, N. Dimitrova, T. McGee, L. Agnihotri, J. Zimmerman, D. Li, and J. Louie, "A probabilistic layered framework fro integrating multimedia content and context information," *Proc. of IEEE Int. Conf. on Acoustics, Speech, and Signal Processing*, vol. 2 , pp. 2057-2060, 2002.

[157] M. R. Naphade and T. S. Huang, "Extracting semantics from audiovisual content: The final frontier in multimedia retrieval," *IEEE Trans. on Neural Networks*, vol. 13, no. 4, pp. 793-810, July 2002.

[158] E. Wold, T. Blum, D. Keislar and J. Wheaton, "Content-based classificaiton, search and retrieval of audio," *IEEE Multimedia*, vol. 3, no. 3, pp. 27-36, 1996.

[159] J. Saunders, Real-Time Discrimination of Broadcast Speech /Music, *IEEE Int. Conf. on Acoustic, Speech, and Signal Processing,*, vol. 2, pp. 993-996, Atlanta, May 1996.

[160] J. Bilmes, "A gentle tutorial on the EM algorithm and its application to parameter estimation for Gaussian mixture and hidden Markov models", *Technical Report ICSI-TR-97-021*, University of Berkeley, 1998.

[161] G. F. Meyer, J. B. Mulligan, and S. M. Wuerger, "Continuous audio-visual digit recognition using N-best decision fusion," *Elsevier International Journal on Multi-Sensor*, Multi-Source Information Fusion, Nov. 2003.

[162] D.W.Massaro, Auditory visual speech processing, in: Proceedings on Eurospeech 2001, Aalborg, 2001, pp. 1153-1156 in G. F. Meyer, J. B. Mulligan, and S. M. Wuerger, "Continuous audio-visual digit recognition using N-best decision fusion," *Elsevier International Journal on Multi-Sensor*, Multi-Source Information Fusion, Nov. 2003.

[163] C. Stauffer, "Automated Audio-Visual Analysis," *MIT Artificial Intelligence Laboratory Memo*, 2005, online: http://people.csail.mit.edu/stauffer/Home.

[164] A.M. Ferman and A.M. Tekalp. "Efficient filtering and clustering methods for temporal video segmentation and visual summarization," *Journal of Visual Comm. and Image Rep.*, 9(4), pp. 336 351, December 1998.

[165] J. Vesanto and E. Alhoniemi, "Clustering of the Self-Organizing Map," *IEEE Transactions on Neural Networks*, 11(3), pp. 586-600, May 2000.

[166] J. R. Smith, S. Basu, C.-Y. Lin, M. Naphade, and B. Tseng, "Interactive content-based retrieval of video," *IEEE Int. Conf. Image Processing*, vol. 1, 2002, pp. 976–979.

[167] J. Vesanto and E. Alhoniemi, "Clustering of the self-organizing map," *IEEE Trans. Neural Network*, vol. 11, no. 3, pp. 586–600, May 2000.

Index

Agent-based automatic relevance feedback (ARF), 116
Anti-reinforcement learning, 156
ARBFN (Adaptive radial basis function network), 57, 61, 93, 94
ARFN (Automatic relevance feedback network), 152, 156, 164
ARFN signal propagation, 154
Audio descriptor, 169
Audio-visual cue, 161, 165, 172
Automatic interaction procedure, 85
Automatic reclusive video retrieval, 149
Automatic Relevance Feedback, 77, 115, 119, 123, 164
Automatic relevance feedback in distributed CBR, 115
AVI (Adaptive video indexing), 129, 133, 134, 162, 165

Brodatz database, 26, 29, 90, 92

CAR (Computer aided referral) system, 42
CBVR (Content-based video retrieval), 129, 161, 165
Centralized storage network, 110
CHSD (Color-histogram-based shot detection), 130, 136
Clustered storage network, 110
CNN news video, 142
Color histogram, 31, 41, 87, 125, 137, 142, 163, 173
Color moments, 31, 41, 125
Community neighborhood discovery, 112
Compressed domain descriptors, 86
Compressed domain image retrieval, 92
Compressed domain processing, 78
Corel photograph collection, 31

Davies-Bonldin index, 163
DCT (Discrete cosine transform), 86
Decision fusion, 172
Distributed digital library, 109
Distributed storage network, 110
DWT (Discrete wavelet transform), 88

EBF (Elliptic basis function), 59
Edge flow model, 120, 121
EDLS (Exact design network using least squares criterion), 62
EM algorithm, 171

Feature fusion, 172
Feature selection method, 90
Fourier descriptor, 31

GW (Gabor wavelet) transformation, 26, 91, 125

Homogeneous texture descriptor, 26
Human Factor, 2
Human-centered search and retrieval, 3

iARM (Interactive-based Analysis and Retrieval of Multimedia), 161
Interactive framework, 13

J2EE (Java 2 enterprise editions), 162
JPEG (Joint photographic experts group), 86
JPEG2000, 88

KFVI (Key frame based video indexing), 130, 132
KFVI (Key frame-based video indexing), 163
Knowledge-based image retrieval, 118

Laplacian mixture model, 169
Lease square (LS) learning, 55
LLN (Law of Large Number), 138
LMN (Local model network), 50
LVQ (Learning vector quantization), 19, 21

Machine learning approach, 49
MCI-CBR (Machine controlled interactive content-based retrieval), 77
MHI (Multiresolution histogram indexing), 89
Mine target detection, 42
MIT texture database, 26
Modified learning vector quantization, 21
Motion feature classification, 132
Movie retrieval, 162, 173
MPEG-4, 88
MPEG-7, 91
Multimedia object, 133

Nonlinear relevance feedback, 15
Nonlinear similarity measure, 12

Object-based video segmentation, 131
Orthogonal least squares (OLS)learning, 55

Peer-to-Peer (P2P) retrieval system, 111

Query and metric adaptation, 14
Query by video clip, 143
Query modification, 13
Query node automatic relevance feedback (ARF), 116
Query packet, 112

RBF (Radial basis function) network, 16, 17, 50, 53, 85, 93, 115, 123
RBF width, 24
Relevance feedback, 13
ROI (Region of interest), 120, 122

Semiautomatic interaction procedure, 85
Shot boundary detection, 136
Side-scan sonar image, 42
Similarity function, 14
Single-class radial basis function network, 17
Single-class RBFN learning, 19
SOM (Self-organizing map), 81
SOTM (Self-organizing tree map), 81, 83, 85, 115, 119, 123
SPIHT, 88

Texture retrieval, 25
TFM (Template-frequency modeling), 139

User subjectivity, 97

Video database, 135
Video group, 135
Video retrieval, 129
Video segmentation, 130
Video story, 135
Visual template, 137

Wavelet moments, 88
WT (Wavelet transform, 86
WT (Wavelet transform), 169

(continued from page ii)

Chaos-Based Digital Communication Systems
Operating Principles, Analysis Methods, and Performance Evalutation
F.C.M. Lau and C.K. Tse
ISBN 3-540-00602-8

Adaptive Signal Processing
Application to Real-World Problems
J. Benesty and Y. Huang (Eds.)
ISBN 3-540-00051-8

Multimedia Information Retrieval and Management Technological
Fundamentals and Applications D. Feng, W.C. Siu, and H.J. Zhang (Eds.)
ISBN 3-540-00244-8

Structured Cable Systems
A.B. Semenov, S.K. Strizhakov,and I.R. Suncheley
ISBN 3-540-43000-8

UMTS
The Physical Layer of the Universal Mobile Telecommunications System
A. Springer and R. Weigel
ISBN 3-540-42162-9

Advanced Theory of Signal Detection
Weak Signal Detection in Generalized Obeservations
I. Song, J. Bae, and S.Y. Kim
ISBN 3-540-43064-4

Wireless Internet Access over GSMand UMTS
M. Taferner and E. Bonek
ISBN 3-540-42551-9